ANALYSIS AND SIMULATION OF NOISE IN NONLINEAR ELECTRONIC CIRCUITS AND SYSTEMS

ANALYSIS AND SIMULATION OF NOISE IN NONLINEAR ELECTRONIC CIRCUITS AND SYSTEMS

by

Alper Demir
Bell Laboratories

and

Alberto Sangiovanni-Vincentelli
University of California

KLUWER ACADEMIC PUBLISHERS
Boston / Dordrecht / London

Distributors for North America:
Kluwer Academic Publishers
101 Philip Drive
Assinippi Park
Norwell, Massachusetts 02061 USA

Distributors for all other countries:
Kluwer Academic Publishers Group
Distribution Centre
Post Office Box 322
3300 AH Dordrecht, THE NETHERLANDS

Library of Congress Cataloging-in-Publication Data

A C.I.P. Catalogue record for this book is available
from the Library of Congress.

Copyright © 1998 by Kluwer Academic Publishers

All rights reserved. No part of this publication may be reproduced, stored in a retrieval system or transmitted in any form or by any means, mechanical, photo-copying, recording, or otherwise, without the prior written permission of the publisher, Kluwer Academic Publishers, 101 Philip Drive, Assinippi Park, Norwell, Massachusetts 02061

Printed on acid-free paper.

Printed in the United States of America

Contents

Acknowledgments		ix
1. INTRODUCTION		1
2. MATHEMATICAL BACKGROUND		5
2.1 Probability and Random Variables		6
	2.1.1 Events and their probabilities	6
	2.1.2 Random variables and their distributions	7
	2.1.3 Expectation	9
	2.1.4 Convergence of random variables	11
	2.1.4.1 Strong law of large numbers	12
	2.1.4.2 Central limit theorem	12
2.2 Stochastic Processes		13
	2.2.1 Mean and autocorrelation	14
	2.2.2 Gaussian processes	15
	2.2.3 Markov processes	16
	2.2.4 Stationary processes	16
	2.2.5 Cyclostationary processes	17
	2.2.6 Spectral density	18
	2.2.7 Wiener process	20
	2.2.8 Poisson process	21
	2.2.9 Continuity and differentiability of stochastic processes	22
	2.2.10 White noise	24
	2.2.11 Ergodicity	26
	2.2.12 Numerical simulation of stochastic processes	30
2.3 Filtering of Stochastic Processes with Linear Transformations		31
	2.3.1 Dynamical system representation	31
	2.3.2 Linear dynamical system representation	32
	2.3.3 Stochastic processes and linear systems	35
	2.3.3.1 WSS processes and LTI systems	35

		2.3.3.2 Cyclostationary processes and LPTV systems	37
2.4	Matrix Algebra, Linear Differential Equations and Floquet Theory		38
	2.4.1	Eigenvalues and eigenvectors of a matrix and its transpose	38
	2.4.2	Similar matrices	39
	2.4.3	Function of a square matrix	40
	2.4.4	Positive definite/semidefinite matrices	41
	2.4.5	Differential equations	42
	2.4.6	Linear homogeneous differential equations	43
	2.4.7	Linear inhomogeneous differential equations	44
	2.4.8	Linear differential equations with constant coefficients	44
	2.4.9	Linear differential equations with periodic coefficients	45
2.5	Stochastic Differential Equations and Systems		48
	2.5.1	Overview	49
	2.5.2	An example	53
	2.5.3	Stochastic integrals	55
	2.5.4	Stochastic differential equations	58
	2.5.5	Ito vs. Stratonovich	60
	2.5.6	Fokker-Planck equation	62
	2.5.7	Numerical solution of stochastic differential equations	63

3. NOISE MODELS — 67
  3.1 Physical Origins of Electrical Noise — 67
    3.1.1 Nyquist's theorem on thermal noise — 70
    3.1.2 Shot noise — 72
    3.1.3 Flicker or $1/f$ noise — 73
  3.2 Model for Shot Noise as a Stochastic Process — 74
  3.3 Model for Thermal Noise as a Stochastic Process — 81
  3.4 Models for Correlated or non-White Noise — 84
    3.4.1 $1/f$ noise model for time-invariant bias — 86
    3.4.2 Models for correlated noise associated with time-varying signals — 89
      3.4.2.1 Probabilistic characterization of the models — 91
      3.4.2.2 Comparison of the models — 93
  3.5 Summary — 96

4. OVERVIEW OF NOISE SIMULATION FOR NONLINEAR ELECTRONIC CIRCUITS — 99
  4.1 Overview — 99
  4.2 Noise Simulation with LTI Transformations — 102
  4.3 Noise Simulation with LPTV Transformations — 106
  4.4 Monte Carlo Noise Simulation with Direct Numerical Integration — 109
  4.5 Summary — 111

5. TIME-DOMAIN NON-MONTE CARLO NOISE SIMULATION — 113
  5.1 Formulation of Circuit Equations with Noise — 114

			Contents	vii

	5.2	Probabilistic Characterization of the Circuit with Noise		117
	5.3	Small Noise Expansion		120
	5.4	Derivation of a Linear Time Varying SDE Model for Noise Analysis		124
	5.5	Derivation of Linear Time Varying ODEs for the Autocorrelation Matrix		128
	5.6	Solution of the Linear Time Varying ODEs for the Autocorrelation Matrix		133
	5.7	Numerical Computation of the Autocorrelation Matrix		133
		5.7.1	Computation of the coefficient matrices	133
		5.7.2	Numerical solution of the differential Lyapunov matrix equation	135
		5.7.3	Numerical solution of the algebraic Lyapunov matrix equation	139
	5.8	Alternative ODEs for the Autocorrelation Matrix		141
		5.8.1	Alternative ODE for the variance-covariance matrix	141
		5.8.2	Alternative ODE for the correlation matrix	142
		5.8.3	Numerical computation of the autocorrelation matrix	143
	5.9	Time-Invariant and Periodic Steady-State		144
		5.9.1	Time-invariant steady-state	144
		5.9.2	Periodic steady-state	146
	5.10	Examples		148
		5.10.1	Parallel RLC circuit	148
		5.10.2	Switching noise of an inverter	150
		5.10.3	Mixer noise figure	150
		5.10.4	Negative resistance oscillator	155
	5.11	Summary		159
6.	NOISE IN FREE RUNNING OSCILLATORS			163
	6.1	Phase Noise and Timing Jitter Concepts		165
	6.2	Phase Noise Characterization with Time Domain Noise Simulation		169
		6.2.1	Definition of timing jitter and phase noise	170
		6.2.2	Probabilistic characterization of phase noise	172
		6.2.3	Examples	176
			6.2.3.1 Ring-oscillator	176
			6.2.3.2 Relaxation oscillator	179
			6.2.3.3 Harmonic oscillator	180
			6.2.3.4 Conclusions	183
		6.2.4	Phase noise "spectrum"	183
	6.3	Phase Noise: Same at All Nodes		185
	6.4	Kaertner's Work on Phase Noise		191
	6.5	Alternative Phase Noise Characterization Algorithm		192
		6.5.1	Examples	194
	6.6	Non-White Noise Sources and Phase Noise		200
	6.7	Phase Noise of Phase-Locked Loops		208
	6.8	Summary		210

7. BEHAVIORAL MODELING AND SIMULATION OF PHASE-LOCKED LOOPS   215
   7.1 PLLs for Clock Generators and Frequency Synthesizers   219
   7.2 Behavioral Models of PLL Components   224
        7.2.1 Reference oscillator and the VCO   224
        7.2.2 Frequency dividers   226
               7.2.2.1 Characterization of timing jitter for frequency dividers 227
        7.2.3 Phase-frequency detector   228
        7.2.4 Charge pump and the loop filter   230
   7.3 Behavioral Simulation Algorithm   232
        7.3.1 Numerical integration with threshold crossing detection 233
        7.3.2 Simulating the timing jitter in numerical integration   236
        7.3.3 Acquisition detection and the simulation output   236
   7.4 Post Processing for Spurious Tones and Timing Jitter/Phase Noise 237
        7.4.1 Spurious tones   238
        7.4.2 Timing jitter/phase noise   240
               7.4.2.1 Variance of the timing jitter of transitions 241
               7.4.2.2 Spectral density of phase noise/timing jitter 245
   7.5 Examples   246
        7.5.1 Acquisition behavior   246
        7.5.2 Timing jitter characterization   249
        7.5.3 Phase noise spectrum   252
   7.6 Summary   256

8. CONCLUSIONS AND FUTURE WORK   261

References   265

Index   273

Acknowledgments

The authors would like to acknowledge the financial support of the Semiconductor Research Corporation (SRC), the California MICRO program (Corporate sponsors: Harris, Philips, Hewlett-Packard, Rockwell, Motorola, Texas Instruments), Bell Laboratories (Lucent Technologies), Cadence Design Systems, Motorola, and CNR (Consiglio Nazionale delle Ricerche) under a VIP grant.

They would like to thank the following individuals for their input and collaboration: Prof. Paul Gray and Prof. Dorit Hochbaum (U.C. Berkeley), Peter Feldmann and Jaijeet Roychowdhury (Bell Laboratories), Ken Kundert (Cadence Design systems), Prof. Jacob White (MIT), Marcel van de Wiel (Philips Research Laboratories), Ed Liu, Iasson Vassiliou, Henry Chang and the other members of the Analog CADgroup at U.C. Berkeley, and Todd Weigandt (U.C. Berkeley).

1 INTRODUCTION

In electronic circuit and system design, the word *noise* is used to refer to any undesired *excitation* on the system. In other contexts, *noise* is also used to refer to signals or excitations which exhibit *chaotic* or *random* behavior. The source of noise can be either *internal* or *external* to the system. For instance, the thermal and shot noise generated within integrated circuit devices are internal noise sources, and the noise picked up from the environment through electromagnetic interference is an external one. Electromagnetic interference can also occur between different components of the same system. In integrated circuits (ICs), signals in one part of the system can propagate to the other parts of the same system through electromagnetic coupling, power supply lines and the IC substrate. For instance, in a mixed-signal IC, the switching activity in the digital parts of the circuit can adversely affect the performance of the analog section of the circuit by traveling through the power supply lines and the substrate.

Prediction of the effect of these noise sources on the performance of an electronic system is called *noise analysis* or *noise simulation*. A methodology for the noise analysis or simulation of an electronic system usually has the following four components:

- Mathematical representations or models for the noise sources.

- Mathematical model or representation for the system that is under the influence of the noise sources.

- A numerical analysis/simulation algorithm to "analyze" or "simulate" the effect of the noise sources on the system in some useful sense.

- Post-processing techniques to *characterize* the effect of the noise sources on the system by calculating useful *performance specifications* using the "data" created by the analysis/simulation of the system.

In this work, we will be concentrating on the type of noise phenomena caused by the small current and voltage fluctuations, such as *thermal*, *shot* and *flicker* noise, which are generated within the integrated-circuit devices themselves. This type of noise is usually referred to as *electrical* or *electronic* noise, because it originates from the fact that electrical charge is not continuous but is carried in discrete amounts equal to the electron charge. Electrical noise is associated with *fundamental* processes in integrated-circuit devices. In practical electronic circuits and systems, the effect of external noise sources, such as digital switching noise coupled through the power supply lines and the IC substrate, can be overwhelming compared with the effect of the electrical noise sources on the performance. The effect of such external noise sources can and should be minimized by using techniques such as differential circuit architectures, separate power supply lines for the analog and digital portions of the circuit, and isolation of the sensitive analog portion from the rest of the system. However, the effect of electrical noise sources can not be eliminated, since it is generated within the electronic devices that make up the system. Thus, electrical noise represents a *fundamental* limit on the performance of electronic circuits [1]. Even though the noise analysis and simulation methodology we will be presenting was developed to analyze the effects of electrical noise, it can also be quite useful in analyzing the effects of other types of noise on electronic circuits. The effects of electrical noise on the performance is most relevant for *analog* and *mixed-signal* electronic circuits, which will be the types of circuits and systems this work is concerned with.

Noise analysis based on *wide-sense stationary* noise source models and the theory of *linear time-invariant* systems has been used by analog circuit and system designers for quite some time. This type of noise analysis, usually referred to as *AC noise analysis*, is implemented in almost every circuit simulator such as SPICE, and it has been quite an invaluable tool in linear analog IC design. For instance, in most cases, analog amplifier circuits operate in small-signal conditions, that is, the "operating-point" of the circuit does not change.

For noise analysis and simulation, the amplifier circuit with a fixed operating-point can be modeled as a linear time-invariant network by making use of the small-signal models of the integrated-circuit devices. On the other hand, some amplifier circuits and other analog circuits such as mixers, filters, oscillators, etc. do not operate in small-signal conditions. For instance, for a mixer circuit, the presence of a large local-oscillator signal causes substantial change in the active devices' operating points over time.

The noise analysis/simulation of electronic circuits and systems that operate in an inherent nonlinear fashion is much more involved than the noise analysis of circuits which can be treated with techniques from the theory of linear time-invariant systems. Even though nonlinear analog circuits has been used extensively in many applications, little work has been done in the analog design and analog CAD community to develop analysis/simulation techniques to characterize the performance of such circuits in the presence of noise sources. This is most probably due to the conceptually difficult nature of the problem, as well as the complexity of the specialized mathematical techniques needed for its treatment. With the explosive growth of the personal mobile communications market, noise analysis/simulation techniques for nonlinear electronic circuits and systems have become an essential part of the design process. Even though most of the signal processing is done in the digital domain, every wireless communication device has an analog front-end which usually is the bottleneck in the design of the whole system. Considerations of power dissipation, form factor and cost push the analog front-end of these devices to higher and higher levels of integration. The requirements for low power operation and higher levels of integration create new challenges in the design of the analog signal processing subsystems of these mobile communication devices. Shrinking dimensions, the push for lower voltage levels, and the use of CMOS technologies for high frequency analog signal processing make the effect of noise on the performance of these inherently nonlinear analog circuits more and more significant. Hence, noise analysis/simulation techniques for the design of such nonlinear analog circuits and systems are bound to be relevant.

In our opinion, some of the work in the literature on the analysis of various noise phenomena in nonlinear electronic circuits and systems treated the problem using ad-hoc and non-rigorous techniques, hence resulting in unreliable and often wrong information about the behavior of the circuit under analysis. We will be treating the problem within the framework of the probabilistic theory of stochastic processes and stochastic differential systems. The well-defined axiomatic structure of the rather well developed theory of probability and stochastic processes provides us a firm foundation in our attempt to understand and analyze the noise phenomena in nonlinear electronic circuits and systems. We would like to point out that a *rigorous* treatment of the problem does not mean that one is dealing with *uncomputable* quantities. Indeed, efficient simulation techniques which have a rigorous foundation can be developed.

We will start in Chapter 2 with a review of the mathematical background needed to follow the rest of our treatment. In Chapter 3, we review the basic physical mechanisms for electrical noise sources and describe their models as stochastic processes. Chapter 4 gives an overview of several noise analysis/simulation techniques that have been proposed in the literature to predict the performance of electronic circuits under the influence of noise sources. In Chapter 5, we describe a novel noise simulation algorithm for nonlinear electronic circuits proposed by us [2, 3]. Then, in Chapter 6, we use the noise simulation algorithm of Chapter 5 to investigate the phase noise/timing jitter phenomenon in free running oscillators. We describe numerical algorithms for the characterization of phase noise in oscillators, and develop models based on its characterization. Top-down hierarchical design methodologies for analog and mixed-signal systems [4] have been proven to be quite effective in dealing with complexity, and in making the design systematic and modular. We believe that a hierarchical approach will be quite effective in dealing with the complexity of the nonlinear noise analysis problem for large and complicated systems. Chapter 7 describes a hierarchical behavioral modeling and simulation methodology for the design of phase-locked loops used in clock generation and frequency synthesis applications. Finally, in Chapter 8, we conclude with a summary and discussion of future directions.

2 MATHEMATICAL BACKGROUND

This chapter presents

- an overview of probability, random variables and stochastic processes,
- filtering of stochastic processes with linear time-invariant, linear time-varying and, in particular, linear periodically time-varying transformations,
- basic results from linear algebra about the eigenvalues and eigenvectors of a matrix and its transpose, positive definiteness, properties of a covariance matrix,
- Floquet theory of a system of linear periodically time-varying differential equations,
- overview of the theory of stochastic differential equations/systems.

Since the topics to be covered are rather extensive, most of the derivations or proofs will not be given here. Definitions and a list of results will be presented as they relate to the analysis of noise in nonlinear systems. For a detailed treatment of the above topics, the reader is referred to the textbooks and other references listed below. We borrowed heavily from these references for our discussion in this chapter.

- Probability, random variables and stochastic processes: [5], [6],[7].
- Filtering of stochastic processes with linear transformations: [5],[8].
- Stochastic differential equations and systems: [9], [10], [11], [12], [13].
- Floquet theory: [14], [15].

2.1 Probability and Random Variables[§]

Mathematical probability has its origins in the games of chance, i.e. *gambling*. Starting in the sixteenth century many famous mathematicians wrote on probability. By the end of the nineteenth century, the lack of a well-defined axiomatic structure was recognized as a serious handicap of the probability theory. In 1933, Andrei Nikolaevich Kolmogorov provided the axioms which are today the foundations for most mathematical probability.

2.1.1 Events and their probabilities

The set of all possible outcomes of an experiment is called the *sample space* and is denoted by Ω. A collection $\mathcal{F}$ of subsets of Ω is called a σ-field if it satisfies the following conditions:

1. $\emptyset \in \mathcal{F}$;
2. if $A_1, A_2, \ldots \in \mathcal{F}$ then $\bigcup_{i=1}^{\infty} A_i \in \mathcal{F}$;
3. if $A \in \mathcal{F}$ then $A^c \in \mathcal{F}$.

A^c denotes the complement of a subset A of Ω. A subset A of Ω is called an *event* if it belongs to the σ-field $\mathcal{F}$.

A *probability measure* P on $(\Omega, \mathcal{F})$ is a function $P : \mathcal{F} \to [0, 1]$ satisfying

1. $P(\emptyset) = 0$, $P(\Omega) = 1$;
2. if $A_1, A_2, \ldots$ is a collection of disjoint members of $\mathcal{F}$, so that $A_i \cap A_j = \emptyset$ for all pairs i, j satisfying $i \neq j$, then

$$P\left(\bigcup_{i=1}^{\infty} A_i\right) = \sum_{i=1}^{\infty} P(A_i). \qquad (2.1)$$

[§]The material in this section is summarized from [5], [6] and [7].

The triple $(\Omega, \mathcal{F}, \mathsf{P})$, comprising a set Ω, a σ-field $\mathcal{F}$ of subsets of Ω, and a probability measure P on $(\Omega, \mathcal{F})$, is called a *probability* space.

Given two events A and B in $\mathcal{F}$, if $\mathsf{P}(B) > 0$ then the *conditional probability* that A occurs given that B occurs is defined to be

$$\mathsf{P}(A \mid B) = \frac{\mathsf{P}(A \cap B)}{\mathsf{P}(B)}. \tag{2.2}$$

A family of events $\{A_i : i \in I, \; A \in \mathcal{F}\}$ is called *independent* if

$$\mathsf{P}\left(\bigcap_{i \in J} A_i\right) = \prod_{i \in J} \mathsf{P}(A_i) \tag{2.3}$$

for all finite subsets J of I.

2.1.2 Random variables and their distributions

A *random variable* is a function $X : \Omega \to \mathbb{R}$ with the property that $\{\omega \in \Omega : X(\omega) \leq x\} \in \mathcal{F}$ for each $x \in \mathbb{R}$. Such an X is called to be $\mathcal{F}$-*measurable*. The *distribution function* of a random variable X is the function $F : \mathbb{R} \to [0, 1]$ given by

$$F(x) = \mathsf{P}(X \leq x) \tag{2.4}$$

where the abbreviation $\{X \leq x\}$ denotes the event $\{\omega \in \Omega : X(\omega) \leq x\} \in \mathcal{F}$.

The random variable X is called *discrete* if it takes values only in some countable subset $\{x_1, x_2, \ldots\}$ of $\mathbb{R}$. The random variable X is called *continuous* if its distribution function can be expressed as

$$F(x) = \int_{-\infty}^{x} f(u)\, du \quad x \in \mathbb{R}, \tag{2.5}$$

for some integrable function $f : \mathbb{R} \to [0, \infty)$. The distribution function of a continuous random variable is obviously continuous. There is also another kind of random variable, called a "singular" random variable. Other random variables (which are not discrete, continuous or singular) are "mixtures" of these three kinds.

The *joint distribution function* of a random vector $\mathbf{X} = [X_1, X_2, \ldots, X_n]$ on the probability space $(\Omega, \mathcal{F}, \mathsf{P})$ is the function $F : \mathbb{R}^n \to [0, 1]$ given by $F(\mathbf{x}) = \mathsf{P}(\mathbf{X} \leq \mathbf{x})$ for $\mathbf{x} = [x_1, x_2, \ldots, x_n] \in \mathbb{R}^n$. The expression $\{\mathbf{X} \leq \mathbf{x}\}$ is an abbreviation for the event $\{\omega \in \Omega : X_1(\omega) \leq x_1, X_2(\omega) \leq x_2, \ldots, X_n(\omega) \leq x_n\} \in \mathcal{F}$. The random variables X and Y on the probability space $(\Omega, \mathcal{F}, \mathsf{P})$ are called *jointly discrete* if the vector $[X, Y]$ takes values only in some countable subset of $\mathbb{R}^2$, and *jointly continuous* if their distribution function can be expressed as

$$F(x,y) = \int_{u=-\infty}^{x} \int_{v=-\infty}^{y} f(u,v) \, du \, dv \qquad x, y \in \mathbb{R}, \tag{2.6}$$

for some integrable function $f : \mathbb{R}^2 \to [0, \infty)$.

The *(probability) mass function* of a discrete random variable X is the function $f : \mathbb{R} \to [0, 1]$ given by $f(x) = \mathsf{P}(X = x)$.

Example 2.1 (Poisson distribution) *If a random variable X takes values in the set $\{0, 1, 2, \ldots\}$ with mass function*

$$f(k) = \frac{\lambda^k}{k!} e^{-\lambda}, \quad k = 0, 1, 2, \ldots \tag{2.7}$$

where $\lambda > 0$, then X is said to have the Poisson distribution *with parameter λ. Poisson distribution plays a key role in modeling shot noise in electronic and other systems.*⋄

The *(probability) density function* of a continuous random variable is the integrable function $f : \mathbb{R} \to [0, \infty)$ in (2.5). The *joint (probability) density function* of two random variables X and Y is the integrable function $f : \mathbb{R}^2 \to [0, \infty)$ in (2.6).

Example 2.2 (Exponential distribution) *The random variable X is exponential with parameter $\lambda > 0$, if it has the density function*

$$F(x) = \lambda e^{-\lambda x}, \quad x \geq 0. \tag{2.8}$$

This distribution is a cornerstone in modeling shot noise, and is closely related to the Poisson distribution. For an exponential random variable X, we have

$$\mathsf{P}(X > s + x \mid X > s) = \mathsf{P}(X > x). \tag{2.9}$$

This is called the lack of memory *property. Exponential distribution is the only continuous distribution with this property.*⋄

Example 2.3 (Gaussian (Normal) distribution) *Probably the most important continuous distribution is the* Gaussian *distribution, which has two parameters μ and σ^2 and the density function*

$$f(x) = \frac{1}{\sqrt{2\pi\sigma^2}} \exp\left(-\frac{(x-\mu)^2}{2\sigma^2}\right), \quad -\infty < x < \infty. \tag{2.10}$$

It is denoted by $N(\mu, \sigma^2)$.⋄

Two random variables X and Y are called *independent* if $\{X \leq x\}$ and $\{Y \leq y\}$ are independent events for all $x, y \in \mathbb{R}$. This definition applies to both discrete and continuous random variables.

Let X and Y be random variables, and let $g, h : \mathbb{R} \to \mathbb{R}$. Then, $g(X)$ and $h(Y)$ are functions which map the sample space Ω into $\mathbb{R}$ by

$$g(X)(\omega) = g(X(\omega)), \quad h(Y)(\omega) = h(Y(\omega)).$$

Let us assume that $g(X)$ and $h(Y)$ are random variables. (This holds only if they are $\mathcal{F}$-measurable.) If X and Y are independent, then so are $g(X)$ and $h(Y)$.

Let X and Y be two jointly continuous random variables on $(\Omega, \mathcal{F}, \mathsf{P})$. The *conditional distribution function* of Y given $X = x$ written $F_{Y|X}(y|x)$ or $\mathsf{P}(Y \leq y | X = x)$, is defined to be

$$F_{Y|X}(y|x) = \int_{v=-\infty}^{y} \frac{f(x,v)}{f_X(x)} dv$$

for any x such that $f_X(x) > 0$. $f_X(x)$ is the density function of X, and $f(x,y)$ is the joint density function of X and Y. Then, the *conditional (probability) density function* of $F_{Y|X}$, written $f_{Y|X}$, is given by

$$f_{Y|X}(y|x) = \frac{f(x,y)}{f_X(x)}.$$

Similarly, conditional distribution functions and conditional density functions can be defined for jointly discrete random variables.

2.1.3 Expectation

Expectation is probably the single most important notion in probability theory. The *expectation* of a discrete random variable X with mass function f is defined to be

$$\mathsf{E}[X] = \sum_{x : f(x) > 0} x \, f(x) \qquad (2.11)$$

whenever this sum is absolutely convergent. The *expectation* of a continuous random variable X with density function f is defined to be

$$\mathsf{E}[X] = \int_{-\infty}^{\infty} x \, f(x) \, dx \qquad (2.12)$$

whenever this integral exists. A rigorous definition of expectation for an arbitrary random variable, regardless of its type (discrete, continuous, and ...), is constructed in measure theory through abstract integration (Lebesgue-Stieltjes integral), which will not be discussed here. The interested reader is referred to the texts on probability and measure theory.

In our treatment, we have first defined probability measure, which then formed the basis for the definition of expectation. It is also possible to first construct a definition for expectation through abstract integration, and then define a probability measure based on the expectation of a special kind of random variable. Hence, expectation is a key concept in probability theory. The expectation operator will be denoted by $\mathbf{E}[\ ]$.

If X and $g(X)$ are continuous random variables then

$$\mathbf{E}\left[g(X)\right] = \int_{-\infty}^{\infty} g(x)\, f(x)\, dx \qquad (2.13)$$

where f is the density function of X. A similar relationship holds for discrete random variables.

If k is a positive integer, then the kth *moment* m_k of a random variable X is defined as

$$m_k = \mathbf{E}\left[X^k\right]. \qquad (2.14)$$

The kth *central moment* σ_k is then defined by

$$\sigma_k = \mathbf{E}\left[(X - m_1)^k\right]. \qquad (2.15)$$

The two moments of most use are $m_1 = \mathbf{E}[X]$ and $\sigma_2 = \mathbf{E}\left[(X - \mathbf{E}[X])^2\right]$, called the *mean* (or *expectation*) and the *variance* of X respectively. The variance of a random variable can not be negative.

Example 2.4 (Mean and variance) *Both the mean and variance of the Poisson distribution defined in Example 2.1 are equal to λ. The mean and variance of the Exponential distribution (Example 2.2) are $1/\lambda$ and $1/\lambda^2$. For the Gaussian distribution (Example 2.3), they are μ and σ^2.* ⋄

Two random variables X and Y are called *uncorrelated* if $\mathbf{E}[XY] = \mathbf{E}[X]\mathbf{E}[Y]$. If X and Y are independent then they are uncorrelated, but the converse is not true in general.

The *covariance* of two random variables X and Y is defined as

$$cov(X, Y) = \mathbf{E}[(X - \mathbf{E}[X])(Y - \mathbf{E}[Y])] = \mathbf{E}[XY] - \mathbf{E}[X]\mathbf{E}[Y]. \qquad (2.16)$$

If X and Y are uncorrelated then $cov(X, Y) = 0$.

A vector $\mathbf{X} = [X_1, X_2, \ldots, X_n]$ of random variables is said to have the *multivariate Gaussian distribution* whenever, for all $\mathbf{a} \in \mathbb{R}^n$, $\mathbf{a}^T \mathbf{X} = a_1 X_1 + a_2 X_2 +$

$\ldots + a_n X_n$ has a Gaussian distribution. Two bivariate Gaussian random variables are independent if and only if they are uncorrelated.

For a vector $\mathbf{X} = [X_1, X_2, \ldots, X_n]$ of random variables, the mean vector is given by $\mu = \mathsf{E}[\mathbf{X}]$, and the $n \times n$ matrix

$$\mathbf{V} = \mathsf{E}\left[(\mathbf{X} - \mu)(\mathbf{X} - \mu)^T\right] \qquad (2.17)$$

is called the *covariance matrix*, because $v_{ij} = cov(X_i, X_j)$.

Let X and Y be two jointly continuous random variables on $(\Omega, \mathcal{F}, \mathsf{P})$. Suppose $X = x$ is given. Conditional upon this, Y has a density function $f_{Y|X}(y|x)$, which is considered to be a function of y. The expected value $\int y f_{Y|X}(y|x) dy$ of this density is called the *conditional expectation* of Y given $X = x$, and is denoted by $\mathsf{E}[Y|X = x]$. The conditional expectation depends on the value x taken by X, and hence can be thought of as a function of X itself. Then the conditional expectation $\mathsf{E}[Y|X]$ is itself a random variable, and one can calculate its expectation. It satisfies the following important property

$$\mathsf{E}[\mathsf{E}[Y|X]] = \mathsf{E}[Y]. \qquad (2.18)$$

2.1.4 Convergence of random variables

The concept of a *limit of a sequence of random variables* naturally arises in many problems in probability theory. Random variables are real-valued functions on some sample space, so convergence of a sequence of random variables is similar to the convergence of a sequence of functions. There are several definitions of convergence for a sequence of functions: pointwise convergence, norm convergence, convergence in measure [6]. These modes of convergence can be adapted to suit families of random variables.

There are four principal ways of interpreting the statement "$X_n \to X$ as $n \to \infty$", where $\{X_n\}$ is a sequence of random variables. Let $X_1, X_2, \ldots, X$ be random variables on some probability space $(\Omega, \mathcal{F}, \mathsf{P})$. We say that

1. $X_n \to X$ *almost surely*, written $X_n \xrightarrow{a.s.} X$ if $\{\omega \in \Omega : X_n(\omega) \to X(\omega)$ as $n \to \infty\}$ is an event whose probability is 1,

2. $X_n \to X$ *in the rth mean*, where $r \geq 1$, written $X_n \xrightarrow{r} X$, if $\mathsf{E}[|X_n^r|] < \infty$ for all n and
$$\mathsf{E}[|X_n - X|^r] \to 0 \text{ as } n \to \infty,$$

3. $X_n \to X$ *in probability*, written $X_n \xrightarrow{P} X$, if
$$\mathsf{P}(|X_n - X| > \varepsilon) \to 0 \text{ as } n \to \infty \text{ for all } \varepsilon > 0,$$

4. $X_n \to X$ *in distribution*, written $X_n \xrightarrow{D} X$, if

$$P(X_n \leq x) \to P(X \leq x) \text{ as } n \to \infty$$

for all points x at which $F(x) = P(X \leq x)$ is continuous.

The definition for almost sure convergence involves a limit for the value of a random variable at a particular point $\omega \in \Omega$ in the sample space: The sequence $\{X_n(\omega)\}$ is a sequence of real numbers, and the limit $X_n(\omega) \to X(\omega)$ as $n \to \infty$ is defined in terms of the definition of limit for a sequence of real numbers. Similarly, the other definitions above involve limits for sequences of probabilities and expectations, which are, basically, sequences of real numbers. These limits are also defined in terms of the limit for a sequence of real numbers. A sequence of real non-random numbers $\{x_n\}$ converges to a number x if and only if for all $\epsilon > 0$, there exists N_ϵ such that $|x_n - x| < \epsilon$ if $n \geq N_\epsilon$.

The following implications hold:

$$(X_n \xrightarrow{a.s.} X) \Rightarrow (X_n \xrightarrow{P} X) \tag{2.19}$$

$$(X_n \xrightarrow{r} X) \Rightarrow (X_n \xrightarrow{P} X) \text{ for any } r \geq 1 \tag{2.20}$$

$$(X_n \xrightarrow{P} X) \Rightarrow (X_n \xrightarrow{D} X) \tag{2.21}$$

$$(X_n \xrightarrow{r} X) \Rightarrow (X_n \xrightarrow{s} X) \text{ for } r > s \geq 1. \tag{2.22}$$

No other implications hold in general.

2.1.4.1 Strong law of large numbers. Let $X_1, X_2, \ldots$ be independent identically distributed (i.i.d.) random variables. Then

$$\frac{1}{n} \sum_{i=1}^{n} X_i \to \mu \text{ almost surely, as } n \to \infty,$$

for some constant μ, if and only if $E[|X_1|] < \infty$. In this case $\mu = E[X_1]$. A sufficient condition for the strong law of large numbers is given by the following: Let $X_1, X_2, \ldots$ be i.i.d. random variables with $E[X_1^2] < \infty$. Then

$$\tfrac{1}{n}\sum_{i=1}^n X_i \xrightarrow{a.s.} \mu, \quad \tfrac{1}{n}\sum_{i=1}^n X_i \xrightarrow{2} \mu \quad \text{as } n \to \infty. \tag{2.23}$$

2.1.4.2 Central limit theorem. Let $X_1, X_2, \ldots$ be independent identically distributed random variables with finite means μ and finite non-zero variances σ^2, and let

$$S_n = X_1 + X_2 + \ldots + X_n.$$

Then
$$\frac{S_n - n\mu}{\sqrt{(n\sigma^2)}} \xrightarrow{D} N(0,1) \text{ as } n \to \infty. \tag{2.24}$$

where $N(0,1)$ denotes the Gaussian distribution with mean 0 and variance 1. The same result also holds under various weaker hypotheses than independence and identical distributions. The central limit theorem is the primary reason why so many random phenomena are modeled in terms of Gaussian random variables. For example, the thermal noise voltage of a resistor is the result of a large number of elementary effects, namely the tiny voltage impulses due to the individual ion cores and free electrons.

2.2 Stochastic Processes[§]

A *stochastic process* X is a family $\{X_t : t \in T\}$ of random variables indexed by some set T and taking values in some set S. There is an underlying probability space $(\Omega, \mathcal{F}, P)$, and each random variable X_t is an $\mathcal{F}$-measurable function which maps Ω into S. For stochastic processes in the study of signals and systems, the index set T usually represents time. We shall only be concerned with the cases when T represents time, and is one of the sets Z, $\{0, 1, 2, \ldots\}$, $\mathbb{R}$, or $[0, \infty)$. When T is an uncountable subset of $\mathbb{R}$, X is called a *continuous-time* stochastic process, and when T is a countable set, such as $\{0, 1, 2, \ldots\}$, it is called a *discrete-time* stochastic process. We shall write $X(t)$ rather than X_t for continuous-time stochastic processes, and X_n for discrete-time processes. The state space S might be a countable (e.g. Z), or an uncountable set such as $\mathbb{R}$ or $\mathbb{R}^n$. We will be mostly dealing with the case when S is an uncountable set, usually $\mathbb{R}$. A stochastic process may be interpreted as a "random function" of time, which is useful in relating evolutionary physical phenomenon to its probabilistic model as a stochastic process, but note that we have not formally defined what a "random function" is. We have formally defined a stochastic process to be a time-indexed family of random variables, and all the mathematical methods and tools of analysis for stochastic processes are developed based on this formal definition.

Evaluation of $X(t)$ at some $\omega \in \Omega$ yields a point in S, which will be denoted by $X(t;\omega)$. For any fixed $\omega \in \Omega$, there is a corresponding collection $\{X(t;\omega) : t \in T\}$ of members of S; this is called a *realization* or *sample path* of the stochastic process X at ω. The "complete" collection of the sample paths of a stochastic process is called the *ensemble*.

The $X(t)$ (for different values of t) are not independent in general. If $S \subseteq \mathbb{R}$ and $\mathbf{t} = [t_1, t_2, \ldots, t_n]$ is a vector of members of T, then the vector

---

[§]The material in this section is summarized from [5], [6] and [7].

$[X(t_1), X(t_2), \ldots, X(t_n)]$ has the joint distribution function $F(\mathbf{x}, \mathbf{t}) : \mathbb{R}^n \to [0, 1]$ given by

$$F(\mathbf{x}, \mathbf{t}) = \mathsf{P}\left(X(t_1) \leq x_1, \ldots, X(t_n) \leq x_n\right).$$

The collection $\{F(\mathbf{x}, \mathbf{t})\}$, as $\mathbf{t}$ ranges over all vectors of members of T of any finite length, is called the collection of *finite-dimensional distributions* (fdds) of X. In general, the knowledge of fdds of a process X does *not* yield complete information about the properties of its sample paths.

Expectation, as defined in Section 2.1.3, can be interpreted as an *ensemble average*, and forms the basis in developing mathematical methods and tools of analysis for stochastic processes in the probabilistic approach to the design and analysis of signals and systems. Expectation in the context of stochastic processes can also be defined in terms of *time averages* instead of ensemble averages. There does exist a *deterministic* theory of stochastic processes based on time averages (i.e. generalized harmonic analysis developed by Norbert Wiener). We choose to use the probabilistic theory of stochastic processes. One important reason for this is that the theory based on ensemble averages (i.e. probabilistic theory) for stochastic processes accommodates *time-varying* averages. On the other hand, time-averages remove all time-varying effects. Thus, to study the statistical behavior of time-varying phenomena, we must rely on the probabilistic theory and models. However the averages measured in practice are often time averages on a single member of an ensemble, that is, a single sample path. For instance, experimental evaluation of *signal-to-noise ratio* (SNR) is often accomplished by time averaging. So, the connection between time averages and ensemble averages needs to be formalized, which is treated in the theory of *ergodicity*.

2.2.1 Mean and autocorrelation

The mean of the stochastic process X, at time t, is simply the mean of the random variable $X(t)$, and is denoted by

$$m_X(t) = \mathsf{E}\left[X(t)\right]. \qquad (2.25)$$

The *autocorrelation* of a stochastic process X, at times t_1 and t_2, is simply the correlation of the two random variables $X(t_1)$ and $X(t_2)$, and is denoted by

$$R_X(t_1, t_2) = \mathsf{E}\left[X(t_1) X(t_2)\right]. \qquad (2.26)$$

The *autocovariance* of a stochastic process X, at times t_1 and t_2, is given by

$$K_X(t_1, t_2) = \mathsf{E}\left[(X(t_1) - m_X(t_1))(X(t_2) - m_X(t_2))\right]. \qquad (2.27)$$

From (2.26) and (2.27), it follows that

$$K_X(t_1, t_2) = R_X(t_1, t_2) - m_X(t_1)m_X(t_2). \quad (2.28)$$

The *cross-correlation* and *cross-covariance* for two random processes X and Y are defined by

$$R_{XY}(t_1, t_2) = \mathbf{E}[X(t_1)Y(t_2)] \quad (2.29)$$
$$K_{XY}(t_1, t_2) = \mathbf{E}[(X(t_1) - m_X(t_1))(Y(t_2) - m_Y(t_2))] \quad (2.30)$$
$$= R_{XY}(t_1, t_2) - m_X(t_1)m_Y(t_2). \quad (2.31)$$

2.2.2 Gaussian processes

A real-valued continuous-time process is called *Gaussian* if each finite dimensional vector $[X(t_1), \ldots, X(t_n)]$ has the multivariate Gaussian distribution $N(\mu(\mathbf{t}), \mathbf{V}(\mathbf{t}))$ for some mean vector μ and some covariance matrix $\mathbf{V}$ which may depend on $\mathbf{t} = [t_1, \ldots, t_n]$.

Gaussian processes are very widely used to model physical phenomena. The central limit theorem, presented in Section 2.1.4.2, is the primary reason for this. If we have a vector of Gaussian processes $\mathbf{X}(t) = [X_1(t), \ldots, X_n(t)]$, the mean vector

$$\mathbf{m_X}(t) = [m_{X_1}(t), \ldots, m_{X_n}(t)] \quad (2.32)$$

and the correlation matrix

$$\mathbf{R_X}(t_1, t_2) = \mathbf{E}\left[\mathbf{X}(t_1)\mathbf{X}(t_2)^T\right] \quad (2.33)$$

for all t_1 and t_2, *completely* specify the fdds for the vector of processes $\mathbf{X}(t)$. This fact together with the central limit theorem leads us to a very important practical conclusion: If we are able to (approximately) model the signals in our system with Gaussian processes (based on the central limit theorem), then to completely characterize these signals probabilistically, all we need is to calculate the *means* and all the *correlations* as given in (2.32) and (2.33). The mean and the correlations are sometimes called the *first* and *second-order probabilistic characteristics* of the process, due to the fact that they are obtained by calculating the first and second-order moments through expectation. Even if the processes are not Gaussian, a second-order probabilistic characterization (i.e. means and correlations) often yields "adequate" information for most practical problems. Hence, means and correlations and methods to calculate them are extremely important from a practical point of view.

2.2.3 Markov processes

The continuous-time process X, taking values in $\mathbb{R}$, is called a *Markov process* if
$$P(X(t_n) \leq x | X(t_1) = x_1, \ldots, X(t_{n-1}) = x_{n-1}) =$$
$$P(X(t_n) \leq x | X(t_{n-1}) = x_{n-1})$$
for all $x, x_1, x_2, \ldots, x_{n-1}$ and all increasing sequences $t_1 < \cdots < t_n$ of times.

A Gaussian process X is a Markov process if and only if
$$E[X(t_n)|X(t_1) = x_1, \ldots, X(t_{n-1}) = x_{n-1}] = E[X(t_n)|X(t_{n-1}) = x_{n-1}]$$
for all $x_1, x_2, \ldots, x_{n-1}$ and all increasing sequences $t_1 < \cdots < t_n$ of times. If a process is both Gaussian and Markov, then its autocovariance function satisfies
$$K_X(t_3, t_1) = \frac{K_X(t_3, t_2) K_X(t_2, t_1)}{K_X(t_2, t_2)}$$
for all $t_1 \leq t_2 \leq t_3$.

Discrete state-space Markov processes are called *Markov chains*, which have been studied extensively in the literature.

2.2.4 Stationary processes

A real-valued stochastic process X is called *nth-order stationary* if its nth-order fdds are invariant under time shifts, that is, if the families
$$\{X(t_1), \ldots, X(t_n)\} \quad \text{and} \quad \{X(t_1 + h), \ldots, X(t_n + h)\}$$
have the same joint distribution for all $t_1, \ldots, t_n$ and $h > 0$. If X is nth-order stationary for every positive integer n, then X is said to be *strictly stationary*. If X is strictly stationary, we must have

$$m_X(t_1) = m_X(t_2) \tag{2.34}$$
$$R_X(t_1, t_2) = R_X(t_1 + h, t_2 + h) \tag{2.35}$$

for all t_1, t_2, and $h > 0$. If it is only known that (2.34) and (2.35) are valid, then X is said to be *wide-sense stationary* (WSS). Similarly, two real-valued processes X and Y are said to be *jointly WSS* if and only if every linear combination $\{aX(t) + bY(t) : a, b \in \mathbb{R}\}$ is a WSS process. It follows that they are jointly WSS if and only if both means, both correlation functions, and the cross-correlation function are invariant under time shifts.

In general, a WSS process is not necessarily strictly stationary. A Gaussian process is strictly stationary if and only if it is WSS.

For a WSS process, the mean m_x is independent of time, and the autocorrelation (and the autocovariance) depends only on the time difference $t_1 - t_2$, i.e.
$$R_X(t_1, t_2) = R_X(t_1 - t_2). \tag{2.36}$$
Hence, for a WSS process X, the autocorrelation is a function of a single variable $\tau = t_1 - t_2$, and is given by
$$R_X(\tau) = \mathrm{E}\left[X(t + \tau/2)X(t - \tau/2)\right]. \tag{2.37}$$
Note that $R_X(\tau)$ in (2.37) is an even function of τ.

2.2.5 Cyclostationary processes

A process X is said to be *wide-sense cyclostationary* if its mean and autocorrelation are periodic with some period T:
$$m_X(t + T) = m_X(t) \tag{2.38}$$
$$R_X(t_1 + T, t_2 + T) = R_X(t_1, t_2) \tag{2.39}$$
for all t, t_1, t_2. The modifier "wide-sense" will be omitted in the rest of our discussion. Unless noted otherwise, we will be dealing with *wide-sense* cyclostationary processes only. With a change of variables $t_1 = t + \tau/2$, $t_2 = t - \tau/2$, we will express the autocorrelation function in (2.39) as
$$R_X(t, \tau) = \mathrm{E}\left[X(t + \tau/2)X(t - \tau/2)\right] \tag{2.40}$$
which is a function of two independent variables, t and τ, is periodic in t with period T for each value of τ, and is an even function of τ. We assume that the Fourier series representation for this periodic function converges, so that R_X in (2.40) can be expressed as
$$R_X(t, \tau) = \sum_{k=-\infty}^{\infty} R_X^{(k)}(\tau) \exp\left(j2\pi k f_c t\right), \tag{2.41}$$
where $f_c = 1/T$ is the fundamental frequency, and the Fourier coefficients $R_X^{(k)}(\tau)$ are given by
$$R_X^{(k)}(\tau) = \frac{1}{T} \int_{-T/2}^{T/2} R_x(t, \tau) \exp\left(-j2\pi k f_c t\right) dt. \tag{2.42}$$
The concept of cyclostationarity can be generalized to *almost cyclostationarity*, in which case the autocorrelation in (2.40) is an *almost periodic function* of t.

A process X is called *mean-square periodic* if

$$\mathsf{E}\left[|X(t+T) - X(t)|^2\right] = 0 \qquad (2.43)$$

for every t and for some period T. From this it follows that, for a specific t

$$\mathsf{P}\left(X(t+T) = X(t)\right) = 1. \qquad (2.44)$$

It does not, however, follow that $X(t;\omega) = X(t+T;\omega)$ for all $\omega \in \Omega$ and for all t. The mean of a mean square periodic process is periodic just like a cyclostationary process. On the other hand, the autocorrelation of a mean square periodic process is *doubly periodic*

$$R_X(t_1 + nT, t_2 + mT) = R_X(t_1, t_2) \qquad (2.45)$$

for all integers m and n. The autocorrelation of a cyclostationary process is not, in general, doubly periodic, because it does not satisfy (2.45) for any m and n, but it does satisfy (2.45) for $m = n$. A mean-square periodic process is cyclostationary, but a cyclostationary process is not necessarily mean-square periodic.

2.2.6 Spectral density

Frequency-domain concepts and methods are widely used in the theory of signals and systems. Fourier series and Fourier transform are the main mathematical tools for frequency-domain analysis. Up to this point in our treatment of stochastic processes, we have not used frequency-domain concepts except for the Fourier series representation of the autocorrelation of a cyclostationary process. At first sight, it is not obvious how we can use Fourier transforms with stochastic processes. One idea is to apply the Fourier transform to the *sample paths* (which was defined before) of stochastic processes. However for most of the stochastic processes (e.g. WSS processes), the sample paths are not finite-energy functions, hence they are not Fourier transformable. A generalized (integrated) Fourier transform was developed by Norbert Wiener in his work on generalized harmonic analysis (1930). We will not further discuss this rather technically involved approach.

Even though the sample paths for stochastic processes are not Fourier transformable in general, it is possible to use Fourier transforms on the autocorrelation functions of stochastic processes and obtain practically useful frequency domain characterizations. The *spectral density* or the *spectrum* of a WSS stochastic process X is defined as the Fourier transform of the autocorrelation function in (2.37), i.e.

$$S_X(f) = \int_{-\infty}^{\infty} R_X(\tau) \exp(-j2\pi f \tau) d\tau \qquad (2.46)$$

whenever the integral exists. Since $R_X(\tau)$ is an even function of τ, $S_X(f)$ is a real and even function. It can also be shown that it is a nonnegative function. From (2.46), we can derive

$$R_X(0) = \mathrm{E}\left[X(t)^2\right] = \int_{-\infty}^{\infty} S_X(f) df \qquad (2.47)$$

using the inverse Fourier transform relationship

$$R_X(\tau) = \int_{-\infty}^{\infty} S_X(f) \exp\left(j2\pi f\tau\right) df. \qquad (2.48)$$

For a nonstationary process X, the autocorrelation is not a function of a single variable in general, i.e.

$$R_X(t,\tau) = \mathrm{E}\left[X(t+\tau/2)X(t-\tau/2)\right]. \qquad (2.49)$$

In this case, an *instantaneous* or *time-varying spectral density* is defined as

$$S_X(t,f) = \int_{-\infty}^{\infty} R_X(t,\tau) \exp\left(-j2\pi f\tau\right) d\tau. \qquad (2.50)$$

For a complete second-order probabilistic characterization of a nonstationary process, specification of either the time-varying autocorrelation function in (2.49) or the time-varying spectral density in (2.50) is needed.

If X is a cyclostationary process, then the time-varying spectral density in (2.50) is periodic in t with fundamental frequency $f_c = 1/T$. We assume that the Fourier series representation for this periodic function converges, so that S_X in (2.50) can be expressed as

$$S_X(t,f) = \sum_{k=-\infty}^{\infty} S_X^{(k)}(f) \exp\left(j2\pi k f_c t\right). \qquad (2.51)$$

It can be shown that

$$S_X^{(k)}(f) = \int_{-\infty}^{\infty} R_X^{(k)}(\tau) \exp\left(-j2\pi f\tau\right) d\tau \qquad (2.52)$$

where $S_X^{(k)}(f)$ are the Fourier series coefficients in (2.51), and $R_X^{(k)}(\tau)$ are the Fourier series coefficients in (2.41). Second-order probabilistic characteristics of cyclostationary processes are completely characterized by either $S_X^{(k)}(f)$ in (2.51) or $R_X^{(k)}(\tau)$ in (2.41).

Nonstationary processes (as opposed to WSS processes) are said to exhibit so-called *spectral correlation*, which is a concept that comes from the deterministic theory of stochastic processes based on time averages. To get an intuitive feeling for spectral correlation, we will discuss the WSS case. This discussion is very non-rigorous and simplistic, but it illustrates the concept of spectral correlation. Let X be a zero-mean WSS process. Assume that the sample paths of the process are Fourier transformable (which is not correct, since a WSS process has sample paths with infinite energy), and the Fourier transform $\bar{X}$

$$\bar{X}(f) = \int_{-\infty}^{\infty} X(t) \exp(-j2\pi ft) dt \qquad (2.53)$$

is also a well-defined zero-mean stochastic process. Then, one can show, by formal manipulation of integrals, that

$$\mathbf{E}\left[\bar{X}(f_1)\bar{X}(f_2)^*\right] = S_X(f_1)\delta(f_1 - f_2), \qquad (2.54)$$

where S_X is the spectral density of X as defined in (2.46), and δ is the Dirac impulse. Hence, the correlation between two different frequency samples of $\bar{X}$ is zero, which would not be true for a nonstationary process.

2.2.7 Wiener process

The Wiener process to be defined in this section is of fundamental importance, not only as a model for a variety of physical phenomena such as Brownian motion, but also as the core of the theory of calculus for stochastic processes. A *standard Wiener process* $W = \{W(t) : t \geq 0\}$ is a real-valued Gaussian process such that

1. $W(0) = 0$.

2. W has *independent stationary increments*:

    - The distribution of $W(t) - W(s)$ depends on $t - s$ alone.
    - The random variables $W(t_j) - W(s_j)$, $1 \leq j \leq n$, are independent whenever the intervals $(s_j, t_j]$ are disjoint.

3. $W(t+s) - W(s)$ is a Gaussian random variable, $N(0, t)$ for all $s, t \geq 0$.

Given the axiomatic definition of the Wiener process, it can be shown that it is a Markov process.

The mean and the autocorrelation function of the Wiener process are given by

$$m_W(t) = \mathbf{E}[W(t)] = 0 \qquad (2.55)$$
$$R_W(t_1, t_2) = \mathbf{E}[W(t_1)W(t_2)] = \min(t_1, t_2). \qquad (2.56)$$

Since a Wiener process is a Gaussian process, the mean and autocorrelation given above completely characterize its fdds. Note that a Wiener process is neither WSS nor cyclostationary.

2.2.8 Poisson process

The Poisson process to be defined in this section is of fundamental importance, not only as a model for shot noise in electronic devices, but also as the core of the theory of point processes. An inhomogeneous *Poisson (counting) process* with time-varying intensity $\lambda(t)$ is a process $N = \{N(t) : t \geq 0\}$ taking values in $S = \{0, 1, 2, \ldots\}$ such that

1. $N(0) = 0$; if $s < t$ then $N(s) \leq N(t)$,

2. $\mathbf{P}\left(N(t+h) = n + m \mid N(t) = n\right) = \begin{cases} \lambda(t)h + o(h) & \text{if } m = 1 \\ o(h) & \text{if } m > 1 \\ 1 - \lambda(t)h + o(h) & \text{if } m = 0 \end{cases}$,

3. if $s < t$ then the number $N(t) - N(s)$ of emissions/crossings/arrivals in the interval $(s, t]$ is independent of the times of emissions/crossings/arrivals during $[0, s]$.

Given the axiomatic definition of the Poisson counting process above, it can be shown that the probability of the number of emissions/crossings/arrivals in the interval (t_1, t_2) being equal to k is given by

$$\mathbf{P}\left(k \text{ emissions in } (t_1, t_2)\right) = \exp\left(-\int_{t_1}^{t_2} \lambda(t) dt\right) \frac{\left(\int_{t_1}^{t_2} \lambda(t) dt\right)^k}{k!}. \tag{2.57}$$

We observe that the number of emissions in the interval (t_1, t_2) is a Poisson random variable (see Example 2.1) with parameter

$$\int_{t_1}^{t_2} \lambda(t) dt.$$

There is an important, alternative and equivalent formulation of a Poisson counting process which provides much insight into its behavior. Let $T_0, T_1, \ldots$ be given by

$$T_0 = 0, \quad T_n = \inf\{t : N(t) = n\}.$$

Then T_n is the time of the nth arrival. The *interarrival times* are the random variables $X_1, X_2, \ldots$ given by

$$X_n = T_n - T_{n-1}.$$

From a knowledge of N, one can determine $X_1, X_2, \ldots$. Conversely, one can reconstruct N from a knowledge of the X_i. For a homogeneous Poisson counting process ($\lambda(t) = \lambda$ a constant), the interarrival times $X_1, X_2, \ldots$ are independent identically distributed exponential random variables with parameter λ. Using the memoryless property of the exponential distribution, one can then show that a homogeneous Poisson process is a Markov process.

The mean and the autocorrelation function of the Poisson process are given by

$$m_N(t) = \mathbf{E}[N(t)] = \int_0^t \lambda(\alpha)d\alpha$$
$$R_N(t_1,t_2) = \int_0^{\min(t_1,t_2)} \lambda(t)dt \left[1 + \int_0^{\max(t_1,t_2)} \lambda(t)dt\right]$$

If $\lambda(t) = \lambda$ is a constant, these reduce to

$$m_N(t) = \lambda t \tag{2.58}$$
$$R_N(t_1,t_2) = \lambda \min(t_1,t_2) + \lambda^2 t_1 t_2. \tag{2.59}$$

Note that the (homogeneous or inhomogeneous) Poisson process is neither WSS nor cyclostationary.

For a homogeneous Poisson process, if it is known that there are exactly k arrivals in an interval (t_1, t_2), then these arrival times have the same statistics as k arbitrary points placed at random in this interval. In other words, the k points can be assumed to be k independent random variables uniformly distributed in the interval (t_1, t_2).

A homogeneous Poisson process has stationary independent increments, meaning that

1. The distribution of $N(t) - N(s)$ depends only on $t - s$.

2. The increments $N(t_j) - N(s_j)$, $1 \leq j \leq n$, are independent, whenever the intervals $(s_j, t_j]$ are disjoint.

2.2.9 Continuity and differentiability of stochastic processes

The notions of *continuity* and *differentiability* for non-random real-valued functions are defined through the concept of limit of sequences of real numbers. Similarly, one can define continuity and differentiability for stochastic processes through the limit of sequences of random variables. In Section 2.1.4, we discussed four different ways of defining convergence for sequences of random variables. Now, we will discuss continuity and differentiability for stochastic processes using mean-square convergence, which can also be defined using other forms of convergence such as almost sure convergence.

A stochastic process X is called *mean-square continuous* at t if

$$\mathrm{E}\left[|X(t) - X(t-h)|^2\right] \to 0 \text{ as } h \to 0. \qquad (2.60)$$

It can be shown that (2.60) is satisfied if and only if the autocorrelation $R_X(t_1, t_2)$ is continuous in t_1 and t_2 at the point $t = t_1 = t_2$, that is, if

$$[R_X(t,t) - R_X(t-h_1, t-h_2)] \to 0 \text{ as } h_1, h_2 \to 0. \qquad (2.61)$$

If X is WSS, then (2.61) is satisfied if and only if $R_X(\tau)$ is continuous at $\tau = 0$, that is, if

$$[R_X(h) - R_X(0)] \to 0 \text{ as } h \to 0 \qquad (2.62)$$

The autocorrelation in (2.56) (for $t_1, t_2 > 0$) for the Wiener process satisfies (2.61) for all $t > 0$. Hence, the Wiener process is mean-square continuous for all $t > 0$.

A stochastic process X is called *mean-square differentiable* at t if there exists a random variable $X'(t)$ such that

$$\mathrm{E}\left[\left|\frac{1}{h}[X(t) - X(t-h)] - X'(t)\right|^2\right] \to 0 \text{ as } h \to 0. \qquad (2.63)$$

If X is mean-square differentiable for all t, then the process $X'(t) = d/dt X(t)$ (assuming that it is a well-defined process in the underlying probability space) is called the *derivative* of X. It can be shown that (2.63) is satisfied if and only if the autocorrelation $R_X(t_1, t_2)$ is differentiable jointly in t_1 and t_2 at the point $t = t_1 = t_2$, that is, if the limit

$$\frac{1}{h_1 h_2}[R_X(t,t) - R_X(t-h_1, t) - R_X(t, t-h_2) + R_X(t-h_1, t-h_2)] \to \left.\frac{\partial^2 R_X(t_1, t_2)}{\partial t_1 \partial t_2}\right|_{t_1 = t_2 = t}$$

as $h_1, h_2 \to 0$

(2.64)

exists. If X is WSS, then the limit in (2.64) exists if and only if $R_X(\tau)$ is twice differentiable at $\tau = 0$, that is, if the limit

$$\frac{1}{h^2}[R_X(h) - 2R_X(0) + R_X(-h)] \to \left.\frac{\partial^2 R_X(\tau)}{\partial \tau^2}\right|_{\tau=0} \text{ as } h \to 0 \qquad (2.65)$$

exists. The autocorrelation in (2.56) for the Wiener process violates (2.64). Hence, the Wiener process is not mean-square differentiable although it is mean-square continuous.

If a process X is mean-square differentiable for all t, and the derivative process X' exists, then the mean and the autocorrelation of the derivative process are given by

$$m_{X'}(t) = \frac{d}{dt}m_X(t) \tag{2.66}$$

$$R_{X'}(t_1, t_2) = \frac{\partial^2 R_X(t_1, t_2)}{\partial t_1 \partial t_2}, \tag{2.67}$$

and if X is WSS, then

$$m_{X'}(t) = 0 \tag{2.68}$$

$$R_{X'}(\tau) = -\frac{\partial^2 R_X(\tau)}{\partial \tau^2}. \tag{2.69}$$

These can be obtained by interchanging the order of the operations of differentiation and expectation (which should be justified).

Mean-square integrability is defined in a similar way through the use of a rectangular Riemann sum.

2.2.10 White noise

The so-called *white noise* is generally understood in engineering literature as a WSS (not necessarily Gaussian) process ξ, for $-\infty < t < \infty$, with mean $\mathbf{E}[\xi(t)] = 0$ and a constant spectral density on the entire real axis, i.e.

$$S_\xi(f) = 1 \text{ for all } f \in \mathbb{R}. \tag{2.70}$$

Using (2.48), we can calculate the autocorrelation function to be

$$R_\xi(\tau) = \mathbf{E}[\xi(t + \tau/2)\xi(t - \tau/2)] = \delta(\tau) \tag{2.71}$$

where δ is the Dirac impulse. From (2.71), we see that ξ is a process such that $\xi(t + \tau/2)$ and $\xi(t - \tau/2)$ are uncorrelated for arbitrarily small values of τ. In particular, we would have

$$\mathbf{E}\left[\xi(t)^2\right] = \int_{-\infty}^{\infty} S_\xi(f) df = \infty, \tag{2.72}$$

which means that the variance of $\xi(t)$ is infinite for all t.

In our above discussion of white noise, we have freely used the Dirac's δ as it is widely used in engineering. As δ is not a well-defined function but a generalized function, white noise is not a well-defined stochastic process. White noise as defined above does not exist as a physically realizable process, and the singular behavior it exhibits does not arise in any realizable context. White noise is a so-called *generalized* stochastic process, and plays a very important role in the theory of calculus for stochastic processes. The differential equations which include white noise as a driving term have to be handled with great care.

If we calculate the autocorrelation function for the derivative W' of the Wiener process using (2.67), we obtain

$$\begin{aligned} R_{W'}(t_1,t_2) &= \frac{\partial^2 R_W(t_1,t_2)}{\partial t_1 \partial t_2} \\ &= \frac{\partial^2}{\partial t_1 \partial t_2} \min(t_1,t_2) \\ &= \delta(t_1 - t_2) \end{aligned} \qquad (2.73)$$

which involves the Dirac impulse. This is the autocorrelation for the WSS white noise process!

Let us define a zero-mean process $\bar{N}$ by subtracting the mean of a homogeneous Poisson process (with parameter λ) from itself:

$$\bar{N}(t) = N(t) - \lambda t.$$

Then the autocorrelation function for $\bar{N}$ is given by

$$R_{\bar{N}}(t_1,t_2) = \lambda \min(t_1,t_2) \qquad (2.74)$$

which is exactly in the same form as the autocorrelation of the Wiener process (except for a constant factor). Hence, the autocorrelation of the derivative $\bar{N}'$ of $\bar{N}$ is given by

$$R_{\bar{N}'}(t_1,t_2) = \lambda \, \delta(t_1 - t_2)$$

which is the autocorrelation of a WSS white noise process (with a constant spectral density equal to λ on the entire real axis). Actually, it can be shown that a process obtained as the formal derivative of any zero-mean process with *stationary* and *uncorrelated increments* has an autocorrelation function in the form of (2.71), and is therefore a white noise process. The WSS white process obtained as the derivative of the Wiener process is a Gaussian process, and therefore called *Gaussian white noise*. The process obtained as the derivative of the Poisson process (after subtraction of the mean) is called *Poisson white noise*. Note that even though both the Wiener process and the Poisson process are not WSS, their "derivative", white noise, is a WSS (generalized) process. White noise is a very useful mathematical idealization for modeling random physical phenomena that fluctuate rapidly with virtually uncorrelated values for different instants of time.

2.2.11 Ergodicity

The theory of *ergodicity* deals with the relationship between ensemble averages (i.e. expectation) and time averages, and hence formalizes the connection between the probabilistic theory of stochastic processes based on ensemble averages, and the deterministic theory based on time averages.

Let X be a stochastic process with mean $m_X(t)$, and the autocorrelation and autocovariance functions $R_X(t,\tau)$ and $K_X(t_1,t_2)$ given by

$$R_X(t,\tau) = \mathrm{E}\left[X(t+\tau/2)X(t-\tau/2)\right] \tag{2.75}$$
$$K_X(t_1,t_2) = \mathrm{E}\left[X(t_1)X(t_2)\right] - m_X(t_1)m_X(t_2) \tag{2.76}$$

We define the *average mean* $\langle m_X \rangle$ for X as the limit

$$\frac{1}{T}\int_{-T/2}^{T/2} m_X(t)dt \to \langle m_X \rangle \text{ as } T \to \infty, \tag{2.77}$$

and the *average autocorrelation* $\langle R_X \rangle(\tau)$ for X as the limit

$$\frac{1}{T}\int_{-T/2}^{T/2} R_X(t,\tau)dt \to \langle R_X \rangle(\tau) \text{ as } T \to \infty \tag{2.78}$$

whenever these limits exist. The *finite-time empirical mean* for a stochastic process X is defined as

$$\hat{m}_X(T) = \frac{1}{T}\int_{-T/2}^{T/2} X(t)dt. \tag{2.79}$$

Note that $\hat{m}_X(T)$ is a random variable (assuming that it is well-defined with the integral above interpreted as the mean-square limit of a rectangular Riemann sum) indexed by T. The *finite-time empirical autocorrelation* for a stochastic process X is defined as

$$\hat{R}_X(\tau)_T = \frac{1}{T}\int_{-T/2}^{T/2} X(t+\tau/2)X(t-\tau/2)dt. \tag{2.80}$$

Note that $\hat{R}_X(\tau)_T$ is a random variable (assuming that it is well-defined with the integral above interpreted as the mean-square limit of a rectangular Riemann sum) indexed by T.

If the following limit

$$\hat{m}_X(T) \xrightarrow{2} \hat{m}_X \text{ as } T \to \infty \tag{2.81}$$

exists (See Section 2.1.4 for the definition of mean-square limit that is denoted by "$\xrightarrow{2}$" in (2.81).), X is said to exhibit *mean-square regularity of the mean*, and $\hat{m}_X$ is called the *empirical mean* of X, which is also a random variable. If the following limit

$$\hat{R}_X(\tau)_T \xrightarrow{2} \hat{R}_X(\tau) \text{ as } T \to \infty \tag{2.82}$$

exists for every τ, X is said to exhibit *mean-square regularity of the autocorrelation*, and $\hat{R}_X(\tau)$ is called the *empirical autocorrelation function* of X, which is also a random variable indexed by τ. It can be shown that mean-square regularity of the mean and the autocorrelation implies that

$$\langle m_X \rangle = \mathbf{E}[\hat{m}_X] \tag{2.83}$$

$$\langle R_X \rangle(\tau) = \mathbf{E}\left[\hat{R}_X(\tau)\right] \tag{2.84}$$

A regular process X is said to have *mean-square ergodicity of the mean* if

$$\mathbf{E}\left[|\hat{m}_X - \mathbf{E}[\hat{m}_X]|^2\right] = 0, \tag{2.85}$$

and a necessary and sufficient condition for this is

$$\frac{1}{T^2} \int_{-T/2}^{T/2} \int_{-T/2}^{T/2} K_X(t_1, t_2) dt_1 dt_2 \to 0 \text{ as } T \to \infty. \tag{2.86}$$

If the process X is WSS, then the condition (2.86) reduces to

$$\frac{1}{T} \int_0^T K_X(\tau) d\tau \to 0 \text{ as } T \to \infty \tag{2.87}$$

where

$$K_X(\tau) = \mathbf{E}[X(t + \tau/2)X(t - \tau/2)] - m_X(t + \tau/2)m_X(t - \tau/2)$$

A regular process X is said to have *mean-square ergodicity of the autocorrelation* if

$$\mathbf{E}\left[\left|\hat{R}_X(\tau) - \mathbf{E}\left[\hat{R}_X(\tau)\right]\right|^2\right] = 0 \tag{2.88}$$

for every τ. A necessary and sufficient condition for the mean-square ergodicity of the autocorrelation can be found by replacing $X(t)$ in (2.86) and (2.87) with $Y_\tau(t) = X(t + \tau/2)X(t - \tau/2)$ where Y_τ is a process which is indexed by τ. Note that

$$m_{Y_\tau}(t) = R_X(t, \tau). \tag{2.89}$$

If X has mean-square ergodicity of the mean and the autocorrelation (which also means that it is regular), then we have

$$\mathrm{E}\left[|\hat{m}_X - \langle m_X \rangle|^2\right] = 0 \tag{2.90}$$

$$\mathrm{E}\left[\left|\hat{R}_X(\tau) - \langle R_X \rangle(\tau)\right|^2\right] = 0. \tag{2.91}$$

For a WSS process X, we have

$$m_X = \mathrm{E}[X(t)] = \langle m_X \rangle \tag{2.92}$$
$$R_X(\tau) = \mathrm{E}[X(t + \tau/2)X(t - \tau/2)] = \langle R_X \rangle(\tau). \tag{2.93}$$

Then, if X is WSS and ergodic (which also means that it is regular), we have

$$\mathrm{E}\left[|\hat{m}_X - m_X|^2\right] = 0 \tag{2.94}$$

$$\mathrm{E}\left[\left|\hat{R}_X(\tau) - R_X(\tau)\right|^2\right] = 0. \tag{2.95}$$

Summarizing:

- For a WSS regular process the empirical autocorrelation function $\hat{R}_X(\tau)$ is in general a random variable (indexed by τ) and its expectation is equal to the probabilistic autocorrelation $R_X(\tau)$. If X is regular, WSS and ergodic then the variance of the empirical autocorrelation is zero. Thus, for a regular, WSS and ergodic X one can calculate the probabilistic autocorrelation by calculating the empirical autocorrelation using a *single* sample path of the process.

- For a nonstationary regular process, the empirical autocorrelation function $\hat{R}_X(\tau)$ is in general a random variable (indexed by τ) and its expectation is equal to the *time-average* of the instantaneous probabilistic autocorrelation $R_X(t, \tau)$ (averaging over t). If X is regular and ergodic then the variance of the empirical autocorrelation is zero. Thus, for a regular, nonstationary, ergodic process X, one can calculate the *time-average* of the instantaneous probabilistic autocorrelation by calculating the empirical autocorrelation using a *single* sample path of the process.

Up to this point, our discussion of ergodicity involved *time-domain* second-order probabilistic characteristics (i.e. the autocorrelation). One can extend the ergodicity concepts for regular processes to *frequency domain* by defining the *empirical spectral density* $\hat{S}_X(f)$ as the Fourier transform of the empirical autocorrelation $\hat{R}_X(\tau)$

$$\hat{S}_X(f) = \int_{-\infty}^{\infty} \hat{R}_X(\tau) \exp(-j2\pi f \tau) d\tau. \tag{2.96}$$

One can define the *average spectral density* $\langle S_X \rangle(f)$ as in (2.78) but, this time, as the time-average of the instantaneous probabilistic spectral density defined by (2.50). Then, for a regular process, expectation of the empirical spectral density is equal to the *time-average* of the instantaneous probabilistic spectral density $S_X(t, f)$ (averaging over t). For a regular and ergodic process the variance of the empirical spectral density is zero. Thus, for a regular, nonstationary, ergodic process X, one can calculate the *time-average* of the instantaneous probabilistic spectral density by calculating the empirical spectral density using a *single* sample path of the process.

For a regular and ergodic process X, although the limit in (2.82) exists and the Fourier transform in (2.96) is well-defined, it can be shown that the limit

$$\lim_{T \to \infty} \frac{1}{T} \int_{-\infty}^{\infty} \left[\int_{-T/2}^{T/2} X(t + \tau/2) X(t - \tau/2) dt \right] \exp(-j2\pi f \tau) d\tau \tag{2.97}$$

does not exist in general in any useful sense (i.e. the order of the limit in (2.82) and the Fourier transform in (2.96) cannot be changed). This has significant practical implications in estimating the empirical spectral density, which we will discuss next.

Let X be a regular and ergodic stochastic process, and let $X_{sp}(t) = X(t; \omega)$ be a sample path for some fixed $\omega \in \Omega$. Our goal is to calculate an estimate for the time-average of the instantaneous probabilistic spectral density for X. Since X is ergodic, we can get this estimate by calculating an estimate of the empirical spectral density using the sample path $X_{sp}(t)$. The *finite-time spectrum* or the *finite-time periodogram* for X is given by

$$\frac{1}{T} |\tilde{X}_T(f)|^2 \tag{2.98}$$

where

$$\tilde{X}_T(f) = \int_{-T/2}^{T/2} X_{sp}(t) \exp(-j2\pi f t) dt. \tag{2.99}$$

It can be shown that (for the reason pointed out above), the limit of the periodogram $1/T |\tilde{X}_T(f)|^2$ as $T \to \infty$ does not exist. It becomes more and more erratic as T is increased. The erratic behavior of (2.98) can be removed by time-averaging before the limit as $T \to \infty$ is taken. This can be accomplished by allowing the location of time interval of analysis $[-T/2, T/2]$ to depend on time u to obtain $[u-T/2, u+T/2]$. The corresponding *time-varying finite-time periodogram* for X is given by

$$\frac{1}{T} |\tilde{X}_T(u,f)|^2 \qquad (2.100)$$

where

$$\tilde{X}_T(u,f) = \int_{u-T/2}^{u+T/2} X_{sp}(t) \exp(-j2\pi ft) dt. \qquad (2.101)$$

Then, it can be shown that

$$\frac{1}{U} \int_{-U/2}^{U/2} \frac{1}{T} |\tilde{X}_T(u,f)|^2 du \to \hat{S}_X(f) \text{ as } T, U \to \infty \qquad (2.102)$$

where $\hat{S}_X(f)$ is the empirical spectral density with mean $\langle S_X \rangle(f)$ and variance equal to zero. In practice, we can not calculate this limit as $T, U \to \infty$, since we only have a finite time-segment of the sample path $X_{sp}(t)$. However, from (2.102), we know that the calculated estimate for the empirical spectral density using (2.102) will get better as we use more information (increase the width of the time interval) for $X_{sp}(t)$. On the other hand, the estimate obtained by calculating (2.98) becomes more and more erratic as T is increased. In practice, we have a sampled version of $X_{sp}(t)$. Hence, the integrals in (2.101) and (2.102) are converted to summations for numerical calculations, and the FFT algorithm is used the calculate the Fourier transform in (2.101).

2.2.12 Numerical simulation of stochastic processes

Effective *analytical methods* for analyzing systems involving stochastic processes are most of the time achievable only in some simple cases. In practice, one is often confronted with the problem of "simulating" a stochastic process on the computer. In this context, *numerical simulation* of a continuous-time stochastic process is meant to mean *generating sample paths* of the process at discrete time points in a specified time interval. Numerical simulation of stochastic processes in this sense, is a vast and widely explored topic.

The most common technique in simulation of stochastic processes is based on a trigonometric representation of stationary stochastic processes. The following assertion is the basis for such approaches: For an arbitrary stationary stochastic process X, for any $\epsilon > 0$ and arbitrary T (sufficiently large), there exist pairwise uncorrelated random variables $A_1, \ldots, A_n$ and $B_1, \ldots, B_n$, and real numbers $f_1, \ldots, f_n$ such that, for arbitrary $t \in [-T, T]$, we have

$$E\left[(X(t) - \sum_{i=1}^{n}(A_k \cos(2\pi f_i t) + B_k \sin(2\pi f_i t))^2)\right] \leq \epsilon. \qquad (2.103)$$

For numerical simulation purposes, one needs to have a way of characterizing or choosing A_k, B_k and f_k. Practical numerical simulations of stationary stochastic processes are most often accomplished by use of finite sum of cosine functions with random phase angles.

We will not further discuss the vast topic of numerical simulation of stochastic processes, but we would like to emphasize the fact that using ad-hoc and "unjustified" ways of simulating stochastic processes will often yield wrong conclusions about the behavior of a system under the effect of the stochastic process. One always has to justify that a representation of a stochastic process, and the approximate numerical simulation scheme based on this representation generate sample paths which converge, in some rigorous and useful sense, to the sample paths of the original stochastic process.

2.3 Filtering of Stochastic Processes with Linear Transformations[§]

In Section 2.2, we have seen that the mean and autocorrelation function/spectral density provide useful information about a stochastic process. We will use stochastic processes to model noise sources. We are naturally interested in investigating the effect of these noise sources as inputs to a "system". We would like to be able to calculate the probabilistic characteristics of the stochastic process at the output of the "system", in particular the mean and the autocorrelation. We will concentrate on linear transformations, because most of the practical systems we will be dealing with can be approximately modeled as linear systems for noise signals. We will investigate the effect of linear transformations on the mean and the autocorrelation function/spectral density of stochastic processes. The effect of the linear transformations on the fdds of the process will not be studied here. It directly follows from the definition of a Gaussian process (see Section 2.2.2) that a linearly transformed Gaussian process is still a Gaussian process, and the fdds for Gaussian processes are completely specified by their mean and autocorrelation functions.

2.3.1 Dynamical system representation

Up to this point, our discussion concentrated on a formalism (theory of probability and stochastic processes) for mathematical representation of "random" signals (i.e. noise signals) as stochastic processes. Now, we will move on to discuss the mathematical representation of systems. We will assume that the systems themselves are deterministic, i.e. "non-random". All systems are also considered to be causal, since they are supposed to model physical systems.

[§]The material in this section is summarized mostly from [5] and [8]. See [5] for a detailed discussion of cyclostationary processes.

A very general *dynamical system representation* as a mathematical model for all kinds of systems can be constructed as a six-tuple which is composed of a set (which usually represents time), an input space (set of input functions), a set of output functions, a set of states, a state-transition function, and a read-out map for the output, along with some axioms about causality, Markovian property, and the semigroup property. This dynamical system representation is general enough to represent differential systems as well as finite-state machines. One can formally define the notions of *linearity* and *time-invariance* based on this dynamical system representation. We will not go into this rather technically involved discussion here. Instead, for our current discussion, we will concentrate on a mathematical representation of a continuous-time dynamical system from an *input/output* relationship perspective, considering it as a black box: A *system* is defined to be a function $\mathcal{H} : \mathbb{C}^{\mathbb{R}} \to \mathbb{C}^{\mathbb{R}}$ that maps an *input* $\{x(t), -\infty < t < +\infty\}$ into an *output* $\{y(t), -\infty < t < \infty\}$, with complex-valued inputs and outputs.[1] In our discussion, we will consider single-input, single-output (SISO) systems for notational simplicity. The extension to multiple-input, multiple-output (MIMO) systems is straightforward.

2.3.2 Linear dynamical system representation

A system $\mathcal{H}$ is said to be *linear* if $\mathcal{H}(ax_1 + bx_2) = a\mathcal{H}(x_1) + b\mathcal{H}(x_2)$, for all a and b in $\mathbb{C}$ and all x_1 and x_2 in $\mathbb{C}^{\mathbb{R}}$. A system $\mathcal{H}$ is said to be *time-invariant* if $\mathcal{H}(\{x(t+\tau), -\infty < t < +\infty\}) = \{\mathcal{H}(x)(t+\tau), -\infty < t < +\infty\}$.

One can define Dirac's $\delta(.)$ by

$$\int_{-\infty}^{\infty} x(u)\delta(t-u)du = x(t) \qquad (2.104)$$

for all $x(.)$ continuous at t. Now, let $\mathcal{H}$ be a linear system. The *impulse response* of $\mathcal{H}$ is defined as $h(t,u) = \mathcal{H}(\delta(t-u))$. The system $\mathcal{H}$ is said to be causal if $h(t,u) = 0$ for $t < u$. It can be shown that

$$\mathcal{H}(x)(t) = \int_{-\infty}^{\infty} x(u)h(t,u)du \qquad (2.105)$$

for some input $x \in \mathbb{C}^{\mathbb{R}}$. If $\mathcal{H}$ is linear and time-invariant (LTI), then $h(t,u) = h(t-u)$.

A linear system $\mathcal{H}$ is said to be *memoryless* if $h(t,u) = c(t)\delta(t-u)$ for some $c : \mathbb{R} \to \mathbb{R}$. An LTI system is memoryless if $h(t-u) = c\delta(t-u)$ for some $c \in \mathbb{R}$.

If the input to an LTI system $\mathcal{H}$ is $x(t) = \exp(j2\pi ft)$, then it can be shown that the output is

$$\mathcal{H}(x)(t) = H(f)\exp(j2\pi ft) \qquad (2.106)$$

where $H(f)$ is the Fourier transform of the impulse response $h(t)$,

$$H(f) = \int_{-\infty}^{\infty} h(t-u) \exp(-j2\pi f(t-u)) du \qquad (2.107)$$

and is called the *system transfer function*. If $x(t) = \int_{-\infty}^{\infty} X(f) \exp(j2\pi ft) df$, i.e. $X(f) = \mathbf{F}\{x(t)\}$, then

$$\mathcal{H}(x)(t) = \int_{-\infty}^{\infty} H(f) X(f) \exp(j2\pi ft) df \qquad (2.108)$$

also

$$Y(f) = \mathbf{F}\{\mathcal{H}(x)(t)\} = H(f) X(f). \qquad (2.109)$$

For a memoryless LTI system with impulse response $h(t) = c\delta(t)$, we have $H(f) = c$, independent of f.

The usefulness of the Laplace/Fourier transform theory for the analysis and synthesis of LTI systems is apparent by looking at (2.105) (an integral relationship) and (2.109) (simple multiplication relationship). The Laplace/Fourier transform belongs to a class of analysis and synthesis techniques called *integral transforms* for linear systems. We will not go into a general discussion of the integral transform theory here. A detailed treatment can be found in [14]. Motivated by the usefulness of the Laplace/Fourier transform theory for LTI systems, an effort was made in the literature to develop a complete operational calculus for linear time-varying (LTV) systems by "extending" the Laplace/Fourier transform theory. Such a transform was proposed by Lötfi Zadeh in 1950 [8]. There was a controversy in the literature on the usefulness of this transform. Nevertheless, for our purposes, this transform forms a rigorous basis for analyzing the effect of noise on LTV systems. By analogy with (2.107), Zadeh defines the *system transfer function* $H(f,t)$ for an LTV system by

$$H(f,t) = \int_{-\infty}^{\infty} h(t,u) \exp(-j2\pi f(t-u)) du. \qquad (2.110)$$

Note that, in contrast to $H(f)$ in (2.107), $H(f,t)$ in (2.110) is a function of both f and t. It can be easily shown that, if the input to an LTV system $\mathcal{H}$ is $x(t) = \exp(j2\pi ft)$, then the output is

$$\mathcal{H}(x)(t) = H(f,t) \exp(j2\pi ft) \qquad (2.111)$$

which is a generalization of (2.106) to LTV systems.

If $x(t) = \int_{-\infty}^{\infty} X(f) \exp(j2\pi ft) df$, i.e. $X(f) = \mathbf{F}\{x(t)\}$, then

$$\mathcal{H}(x)(t) = \int_{-\infty}^{\infty} H(f,t) X(f) \exp(j2\pi ft) df. \qquad (2.112)$$

For a memoryless LTV system with $h(t, u) = c(t)\delta(t-u)$, we have $H(f, t) = c(t)$, independent of f.

A linear system is said to be (linear) periodically time-varying (LPTV), if the impulse response satisfies

$$h(t, u) = h(t + T, u + T) \tag{2.113}$$

for all $t, u \in \mathbb{R}$, and for some period $T > 0$. Then, for an LTPV system, the impulse response $h(t + \tau, t)$ is periodic in t for every τ, and can therefore be represented by a Fourier series (assumed to converge)

$$h(t + \tau, t) = \sum_{n=-\infty}^{\infty} h_n(\tau) \exp(j2\pi n f_c t) \tag{2.114}$$

where $f_c = 1/T$ is the fundamental frequency, and the Fourier coefficients $h_n(\tau)$ are given by

$$h_n(\tau) = \frac{1}{T} \int_{-T/2}^{T/2} h(t + \tau, t) \exp(-j2\pi n f_c t) dt. \tag{2.115}$$

Then the system transfer function

$$H(f, t) = \int_{-\infty}^{\infty} h(t, t - \tau) \exp(-j2\pi f \tau) d\tau \tag{2.116}$$

is also periodic in t and can be represented by a Fourier series

$$H(f, t) = \sum_{n=-\infty}^{\infty} H_n(f + nf_c) \exp(j2\pi n f_c t) \tag{2.117}$$

where

$$H_n(f) = \int_{-\infty}^{\infty} h_n(\tau) \exp(-j2\pi f \tau) d\tau. \tag{2.118}$$

If the input to an LPTV system $\mathcal{H}$ is $x(t) = \exp(j2\pi ft)$, then the output is

$$\mathcal{H}(x)(t) = H(f, t) \exp(j2\pi ft) = \sum_{n=-\infty}^{\infty} H_n(f + nf_c) \exp(j2\pi(f + nf_c)t) \tag{2.119}$$

which is a special case of (2.111). If $x(t) = \int_{-\infty}^{\infty} X(f) \exp(j2\pi ft) df$ is the input to an LPTV system, i.e. $X(f) = \mathbf{F}\{x(t)\}$, then the output is given by

$$\begin{aligned}
\mathcal{H}(x)(t) &= \int_{-\infty}^{\infty} H(f, t) X(f) \exp(j2\pi ft) df \\
&= \int_{-\infty}^{\infty} \sum_{n=-\infty}^{\infty} H_n(f + nf_c) \exp(j2\pi n f_c t) X(f) \exp(j2\pi ft) df \\
&= \mathbf{F}^{-1} \left\{ \sum_{n=-\infty}^{\infty} H_n(f) X(f + nf_c) \right\}.
\end{aligned} \tag{2.120}$$

Hence, the Fourier transform of the output is $Y(f) = \sum_{n=-\infty}^{\infty} H_n(f) X(f + nf_c)$, which is a generalization of (2.109) to LPTV systems. For a memoryless LPTV system with $h(t,u) = c(t)\delta(t-u)$, we have $H(f,t) = c(t)$, independent of f, and $c(t)$ is periodic in t.

From (2.106), we observe that if a single complex exponential at frequency f is input to an LTI system, the output is also a single complex exponential at frequency f with a scaled amplitude, where the scaling is set by the transfer function $H(f)$. For an LTV system, the output for a single complex exponential input, in general, contains a continuum of frequencies. For LPTV systems, from (2.119), we observe that the output corresponding to a single complex exponential at frequency f is a summation of complex exponentials at frequencies $f + nf_c$, $n \in \mathbb{Z}$, where f_c is the fundamental frequency for the LPTV system. It is interesting to compare this with the fact that the *steady-state* response of a *nonlinear time-invariant* system to a complex exponential with frequency f is a summation of complex exponentials at the harmonics of the input frequency f.

2.3.3 Stochastic processes and linear systems

In the previous section, we considered deterministic inputs to linear systems. Now, let a stochastic process X be the input to a linear system with impulse response $h(t, u)$. Note that the system is considered to be deterministic. Then the output is another stochastic process Y given by

$$Y(t) = \int_{-\infty}^{\infty} X(u) h(t, u) du \qquad (2.121)$$

assuming that the above integral is well-defined for all t in some rigorous sense (e.g. mean-square integrability). Then the mean and the autocorrelation of Y, in terms of the mean and autocorrelation of X, are given by

$$m_Y(t) = \int_{-\infty}^{\infty} m_X(u) h(t, u) du \qquad (2.122)$$

$$R_Y(t_1, t_2) = \int_{-\infty}^{\infty} \int_{-\infty}^{\infty} R_X(r, s) h(t_1, r) h(t_2, s) dr ds \qquad (2.123)$$

2.3.3.1 WSS processes and LTI systems. If the input X is a WSS process and the system is a *stable* LTI system (with impulse response $h(t)$), then it can be shown that the output Y is also a WSS process. A sufficient condition for stability is

$$\int_{-\infty}^{\infty} h(t)^2 dt < \infty. \qquad (2.124)$$

For instance, an ideal integrator does not satisfy the stability condition. When a WSS process is the input to an integrator, the output is, in general, *not* a WSS process.

Assuming that the LTI system is stable and the output is WSS, then the mean and autocorrelation of the output are given by

$$m_Y = m_X \int_{-\infty}^{\infty} h(t)dt \qquad (2.125)$$

and

$$R_Y(\tau) = \int_{-\infty}^{\infty} R_X(u) r_h(\tau - u) du \qquad (2.126)$$
$$= R_X(\tau) \odot r_h(\tau) \qquad (2.127)$$

where $r_h(\tau)$ is given by

$$r_h(\tau) = h(\tau) \odot h(-\tau) \qquad (2.128)$$

and $\odot$ denotes the convolution operation. The cross-correlations of the input X and output Y are given by

$$R_{YX}(\tau) = R_X(\tau) \odot h(\tau), \qquad (2.129)$$
$$R_{XY}(\tau) = R_X(\tau) \odot h(-\tau). \qquad (2.130)$$

Then, if we assume that the autocorrelation of the output possesses a Fourier transform, the spectral density of the output is given by

$$S_Y(f) = |H(f)|^2 S_X(f) \qquad (2.131)$$

where $H(f)$ is the transfer function. This relationship plays a crucial role in the analysis of signal processing systems involving linear time-invariant transformations, and is the basis for analyzing the effect of noise which can be modeled as a WSS process input to an LTI system. (2.131) also shows how the spectral density links to the notion of a *frequency decomposition* of the power in the stochastic process X. Recall that the total power for the output process Y is given by

$$\mathrm{E}\left[Y(t)^2\right] = \int_{-\infty}^{\infty} |H(f)|^2 S_X(f) df \qquad (2.132)$$

which is a constant as a function of t, since Y is WSS.

The *cross-spectral densities* for the input and the output, defined as the Fourier transforms of the cross-correlations, are given by

$$S_{YX}(\tau) = S_X(f) H(f), \qquad (2.133)$$
$$S_{XY}(\tau) = S_X(f) H(-f). \qquad (2.134)$$

The spectral density is an even and real function, but cross-spectral densities are, in general, not real-valued functions. Cross-spectral densities play a crucial role in evaluating the degree to which two processes are approximately related by an LTI transformation. The relationship (2.133) is important from a system identification point of view: The transfer function of an LTI system can be calculated easily if we can calculate the cross- spectral density of the input and the output. We can choose any input process with a known spectral density $S_X(f)$.

2.3.3.2 Cyclostationary processes and LPTV systems. Another special case of interest is when cyclostationary processes are inputs to LPTV systems. Assuming that the cyclostationary input process X has the same fundamental frequency as the LPTV system, it can be shown that the output process Y is also cyclostationary with the same fundamental frequency, provided that the LPTV system satisfies a "stability" condition. This condition will be discussed in detail later in the chapter, now we concentrate on relating the autocorrelation function/spectral density of the output to those of the input through the transfer function of the LPTV system. The Fourier series coefficients given by (2.42)/(2.52), and (2.115)/(2.118) completely characterize the autocorrelation function/time-varying spectral density for a cyclostationary process, and the impulse response/transfer function for an LPTV system respectively. It can be shown that the Fourier series coefficients for the autocorrelation of the cyclostationary process Y at the output are given by

$$R_Y^{(k)}(\tau) = \sum_{n,m=-\infty}^{\infty} R_X^{(k-(n-m))}(\tau) \exp\left(-j\pi(n+m)f_c\tau\right) \odot r_{nm}^{(k)}(-\tau) \quad (2.135)$$

where

$$r_{nm}^{(k)}(\tau) = \int_{-\infty}^{\infty} h_n(t+\tau/2) h_m^*(t-\tau/2) \exp\left(-j2\pi k f_c t\right) dt. \quad (2.136)$$

Then, it can be shown that the Fourier series coefficients for the time-varying spectral density of the output are given by

$$S_Y^{(k)}(f) = \sum_{n,m=-\infty}^{\infty} H_n(f+kf_c/2) S_X^{(k-(n-m))}(f-(n+m)f_c/2) H_m^*(f-kf_c/2).$$
$$(2.137)$$

We might be interested in some special cases. For instance, if the system is LTI and the input is cyclostationary, (2.135) and (2.137) reduce to

$$R_Y^{(k)}(\tau) = R_X^{(k)}(\tau) \odot r_h^{(k)}(-\tau) \quad (2.138)$$

and
$$S_Y^{(k)}(f) = H(f + kf_c/2)S_X^{(k)}(f)H^*(f - kf_c/2) \qquad (2.139)$$
where
$$r_h^{(k)}(\tau) = \int_{-\infty}^{\infty} h(t + \tau/2)h(t - \tau/2)\exp(-j2\pi kf_c t)dt \qquad (2.140)$$

and, $h(t)$ is the impulse response, and $H(f)$ is the transfer function. Another special case is when the input is WSS and the system is LPTV. In this case, the output, is in general, cyclostationary. Then, (2.135) and (2.137) reduce to

$$R_Y^{(k)}(\tau) = \sum_{n=-\infty}^{\infty} R_X(\tau)\exp(-j\pi(2n-k)f_c\tau) \odot r_{n(n-p)}^{(k)}(-\tau) \qquad (2.141)$$

and

$$S_Y^{(k)}(f) = \sum_{n=-\infty}^{\infty} H_{n+k}(f + kf_c/2)S_X(f - (k/2+n)f_c)H_n^*(f - kf_c/2). \qquad (2.142)$$

One can also calculate the cross-correlations and cross-spectral densities for the input and output processes. Cross-spectral densities play a crucial role in the identification of LPTV systems. A generalization of (2.133) holds for cyclostationary processes and LPTV systems.

All the above results for cyclostationary processes and LPTV systems can be generalized for *almost* cyclostationary processes and linear *almost* periodically time-varying systems.

2.4 Matrix Algebra, Linear Differential Equations and Floquet Theory[§]

Now, we will cover some basic results from matrix algebra and the differential equation theory.

2.4.1 Eigenvalues and eigenvectors of a matrix and its transpose

Given a real $n \times n$ matrix $\mathbf{A}$, the polynomial $d(\lambda) = \det(\lambda \mathbf{I} - \mathbf{A})$ is called the *characteristic polynomial* of the matrix $\mathbf{A}$, and the equation $d(\lambda) = 0$ is called the *characteristic equation* of $\mathbf{A}$. Then, $d(\lambda)$ is an n-th degree monic polynomial with real coefficients. Thus, the characteristic equation has n roots, which are called the *eigenvalues* of the matrix $\mathbf{A}$. If λ is an eigenvalue of $\mathbf{A}$, then $\lambda \mathbf{I} - \mathbf{A}$ is singular, and the equation

$$(\lambda \mathbf{I} - \mathbf{A})\mathbf{x} = 0 \qquad (2.143)$$

[§]See [14], [15] and textbooks on linear algebra and linear system theory for proofs and more details. Some of the material in this section is summarized from the notes of the linear system theory course taught by Prof. Erol Sezer in Fall 1990 at Bilkent University, Turkey.

MATHEMATICAL BACKGROUND 39

has nontrivial solutions. Each of these solutions is called an *eigenvector* of **A** corresponding to the eigenvalue λ. Eigenvalues of a real matrix may be real or complex, however complex eigenvalues occur in conjugate pairs. If λ and λ^* are a pair of complex conjugate eigenvalues, and if **x** is an eigenvector corresponding to λ, then $\mathbf{x}^*$ is an eigenvector corresponding to λ^*. A matrix can have repeated eigenvalues. A square matrix is nonsingular if and only if it does not have any zero eigenvalues.

It can be easily shown that the matrix **A** and its transpose $\mathbf{A}^T$ have the same eigenvalues. Let us assume that the real $n \times n$ matrix **A**, and hence its transpose $\mathbf{A}^T$, have *distinct* eigenvalues $\lambda_1, \lambda_2, \ldots, \lambda_n$. Let $\mathbf{x}_1, \mathbf{x}_2, \ldots, \mathbf{x}_n$ be the corresponding eigenvectors for **A**, and $\mathbf{y}_1, \mathbf{y}_2, \ldots, \mathbf{y}_n$ be the corresponding eigenvectors for $\mathbf{A}^T$. Then, we have

$$\lambda_i \mathbf{y}_j^T \mathbf{x}_i = \mathbf{y}_j^T \lambda_i \mathbf{x}_i \qquad (2.144)$$

$$= \mathbf{y}_j^T \mathbf{A} \mathbf{x}_i \qquad (2.145)$$

$$= \lambda_j \mathbf{y}_j^T \mathbf{x}_i. \qquad (2.146)$$

If $i = j$, the above reduces to a trivial equality, but for $i \neq j$ we have

$$\lambda_i \mathbf{y}_j^T \mathbf{x}_i = \lambda_j \mathbf{y}_j^T \mathbf{x}_i. \qquad (2.147)$$

For $i \neq j$, (2.147) is satisfied only if $\mathbf{y}_j^T \mathbf{x}_i = 0$, since $\lambda_i \neq \lambda_j$. It can also be shown that for $i = j$, we must have $\mathbf{y}_j^T \mathbf{x}_i \neq 0$. So, we conclude that, for a matrix with distinct eigenvalues, the eigenvectors of the matrix are *orthogonal* to the eigenvectors of the transposed matrix corresponding to different eigenvalues, i.e.

$$\mathbf{y}_j^T \mathbf{x}_i \begin{cases} = 0 & \text{if } i \neq j \\ \neq 0 & \text{if } i = j \end{cases} \qquad (2.148)$$

The eigenvalues of a real symmetric matrix are real. If the eigenvalues of a matrix **A** are $\lambda_1, \ldots, \lambda_n$, then the eigenvalues of the matrix $c\mathbf{I} + \mathbf{A}$ are $c + \lambda_1, \ldots, c + \lambda_n$, where $c \in \mathbb{R}$.

2.4.2 Similar matrices

Two matrices **A** and **B** which are related as $\mathbf{B} = \mathbf{U}^{-1}\mathbf{A}\mathbf{U}$, where **U** is a nonsingular matrix, are called *similar*, and the process of obtaining **B** from **A** is called a *similarity transformation*. Similar matrices have the same characteristic polynomial, and therefore, the same eigenvalues.

It can be shown that the eigenvectors of a matrix corresponding to *distinct* eigenvalues are linearly independent. If an $n \times n$ matrix **A** has simple eigenvalues, that is, if the eigenvalues are distinct, then the matrix **A** is similar to a diagonal matrix whose diagonal values are the eigenvalues of **A**. Let $\lambda_1, \lambda_2, \ldots, \lambda_n$ be the simple eigenvalues, and $\mathbf{x}_1, \mathbf{x}_2, \ldots, \mathbf{x}_n$ be the corresponding eigenvectors. Then, the matrix

$$\mathbf{P} = [\mathbf{x}_1, \mathbf{x}_2, \ldots, \mathbf{x}_n]$$

is nonsingular, and

$$\mathbf{P}^{-1}\mathbf{A}\mathbf{P} = \text{diag.}(\lambda_1, \lambda_2, \ldots, \lambda_n).$$

The matrix **P**, which consists of the eigenvectors of **A**, is called the *modal matrix* of **A**. Note that, neither the eigenvectors nor the modal matrix are unique, but every modal matrix reduces **A** to its diagonal form.

Every matrix with simple eigenvalues can be diagonalized by a similarity transformation. In the general case when the matrix has multiple eigenvalues, whether the matrix can be diagonalized depends on whether or not we can find sufficient number of linearly independent eigenvectors for each eigenvalue. Some matrices with multiple eigenvalues can not be diagonalized. However, they can still be reduced to a simple form, not diagonal though, by means of a similarity transformation, which is called the *Jordan* form. We will not further discuss this here.

2.4.3 Function of a square matrix

Let **A** be an $n \times n$ matrix, and let $p(\lambda) = a_0 + a_1\lambda + \ldots + a_m\lambda^m$ be a polynomial in the variable λ with real coefficients. The polynomial $p(\mathbf{A})$ of the matrix **A** is defined as

$$p(\mathbf{A}) = a_0\mathbf{I} + a_1\mathbf{A} + \ldots + a_m\mathbf{A}^m. \qquad (2.149)$$

It can be shown that every matrix satisfies its own characteristic polynomial. This result is known as the *Cayley-Hamilton* theorem.

Let a matrix **A** have distinct eigenvalues $\lambda_1, \ldots, \lambda_k$ with multiplicities $n_1, \ldots, n_k$ in the characteristic polynomial. Let $f(\lambda)$ be a function which is analytic in some simply connected region of the complex plane that contains $\lambda_1, \ldots, \lambda_k$. Then, there exists a unique polynomial $p(\lambda)$ of degree $n-1$ where $n = \sum_{i=1}^{k} n_i$, such that for each i

$$p^{(j)}(\lambda_i) = f^{(j)}(\lambda_i), \quad j = 1, 2, \ldots, n_i$$

where $p^{(j)}(\lambda_i)$ denotes the j-th derivative of p evaluated at $\lambda = \lambda_i$. Then the function $f(\mathbf{A})$ of the matrix **A** is defined to be $p(\mathbf{A})$. The polynomial $p(\lambda)$ is called the *annihilating polynomial* for $f(\lambda)$.

It can be shown that if the matrix **A** has distinct eigenvalues $\lambda_1, \lambda_2, \ldots, \lambda_k$ with multiplicities $n_1, n_2, \ldots, n_k$, and if the function $f(\mathbf{A})$ is defined, then the matrix $f(\mathbf{A})$ has the eigenvalues $f(\lambda_1), f(\lambda_2), \ldots, f(\lambda_k)$ with multiplicities $n_1, n_2, \ldots, n_k$.

If **A** is diagonalizable, i.e. there exists nonsingular **P** such that $\mathbf{A} = \mathbf{PDP}^{-1}$, where $\mathbf{D} = \text{diag.}(\lambda_i)$, then

$$f(\mathbf{A}) = \mathbf{P}\text{diag.}(f(\lambda_i))\mathbf{P}^{-1}. \qquad (2.150)$$

2.4.4 Positive definite/semidefinite matrices

A symmetric $n \times n$ matrix **A** ($\mathbf{A} = \mathbf{A}^T$) is said to be

1. positive definite, if $\mathbf{x}^T \mathbf{A} \mathbf{x}$ has a positive value for all $\mathbf{x} \neq \mathbf{0}$,

2. positive semidefinite, if $\mathbf{x}^T \mathbf{A} \mathbf{x} = 0$ for some $\mathbf{x} \neq \mathbf{0}$, and has a positive value for the remaining $\mathbf{x} \neq \mathbf{0}$,

3. indefinite, if $\mathbf{x}^T \mathbf{A} \mathbf{x}$ takes both positive and negative values.

$q(\mathbf{x}) = \mathbf{x}^T \mathbf{A} \mathbf{x}$ is called a *quadratic form*. It can be shown that a symmetric matrix **A** is positive definite if and only if all of its eigenvalues are positive, and is positive semidefinite if and only if at least one eigenvalue is zero and the remaining ones are positive. Note that the eigenvalues of any real symmetric matrix are real. It can easily be shown that the diagonal elements of a positive semidefinite matrix are nonnegative, and the diagonal elements of a positive definite matrix are strictly positive.

In Section 2.1.3, we have defined the covariance matrix **V** for a vector of random variables **X** as

$$\mathbf{V} = \mathrm{E}\left[(\mathbf{X} - \mu)(\mathbf{X} - \mu)^T\right] \qquad (2.151)$$

By definition, the covariance matrix **V** is a symmetric matrix. Now, let us define the random variable $Y = \mathbf{a}^T \mathbf{X}$, where $\mathbf{a} \in \mathbb{R}^n$. Then, the variance of Y is given by

$$var(Y) = \mathbf{a}^T \mathbf{V} \mathbf{a}. \qquad (2.152)$$

We know that $var(Y) \geq 0$ for a random variable Y, hence we conclude that the covariance matrix **V** for a vector of random variables **X** is a positive semidefinite matrix. If there exists an $\mathbf{a} \neq \mathbf{0}$ such that $\mathbf{a}^T \mathbf{V} \mathbf{a} = 0$, then the variance of the random variable $Y = \mathbf{a}^T \mathbf{X}$ is zero.

Let **A** be a real symmetric matrix. Then, it can be shown that there exists an orthogonal matrix **P** ($\mathbf{P}^T \mathbf{P} = \mathbf{P}\mathbf{P}^T = \mathbf{I}$) such that

$$\mathbf{P}^T \mathbf{A} \mathbf{P} = \mathbf{D} = diag.(\lambda_i) \qquad (2.153)$$

where **D** is a diagonal matrix with eigenvalues of **A** as the diagonal elements.

For the covariance matrix of the vector of random variables **X**, the eigenvalues are nonnegative since it is a semidefinite matrix. Then, we have

$$\mathbf{U}^T\mathbf{V}\mathbf{U} = \mathbf{D} = diag.(\lambda_i) \quad (2.154)$$

for some orthogonal matrix **U**, where **D** is a diagonal matrix with nonnegative diagonal elements. Let us define the vector of random variables $\mathbf{Z} = \mathbf{U}^T\mathbf{X}$. Then, it can be shown that the covariance matrix of **Z** is **D**. Since **D** is diagonal, the entries of **Z** are pairwise uncorrelated. Note that, we also have $\mathbf{X} = \mathbf{U}\mathbf{Z}$, since **U** is an orthogonal matrix.

Let **A** be a real $n \times n$ symmetric matrix. **A** is positive definite if and only if $\mathbf{A} = \mathbf{B}^T\mathbf{B}$ for some nonsingular $n \times n$ matrix **B**. **A** is positive semidefinite if and only if $\mathbf{A} = \mathbf{C}\mathbf{C}^T$ for some $n \times r$ matrix **C**, where $r < n$. If $r > n$, **A** is positive definite or positive semidefinite depending on the rank of **C**.

2.4.5 Differential equations

Let

$$\dot{\mathbf{x}} = \mathbf{f}(t,\mathbf{x}) \quad \mathbf{x}(t_0) = \mathbf{x}_0 \quad t \in \mathbb{R}, \; \mathbf{x} \in \mathbb{R}^n, \; \mathbf{f}: \mathbb{R} \times \mathbb{R}^n \to \mathbb{R}^n \quad (2.155)$$

If the following conditions are satisfied

1. For every fixed **x**, the function $t \to \mathbf{f}(t,\mathbf{x})$ ($\mathbf{f}(.,\mathbf{x})$) is piecewise continuous.

2. **f** satisfies a Lipschitz condition, i.e. there exists a piecewise continuous $k(.): \mathbb{R} \to \mathbb{R}_+$ such that

$$\|\mathbf{f}(t,\mathbf{x}) - \mathbf{f}(t,\mathbf{x}')\| \leq k(t)\|\mathbf{x} - \mathbf{x}'\| \quad \forall \mathbf{x}, \mathbf{x}' \in \mathbb{R}^n$$

which is a stronger condition than continuity. For instance $f(x) = \sqrt{x}$ is continuous, but does not satisfy a Lipschitz condition.

then, the differential equation in (2.155) has a *unique* continuous solution. The proof of the existence of the solution is done by defining a sequence $\{\mathbf{x}_m\}_{m=0}^{\infty}$ in the space of continuous functions defined on an interval $[t_0, t_1]$ (which is a complete vector space, meaning that every Cauchy sequence in this vector space converges to a vector in the space) as $\mathbf{x}_0(t) = \mathbf{x}_0$, and

$$\mathbf{x}_m(t) = \mathbf{x}_0 + \int_{t_0}^{t} \mathbf{f}(\tau, \mathbf{x}_{m-1}(\tau))d\tau \quad m = 1, 2, \ldots \quad (2.156)$$

It can be shown that this sequence converges to a continuous function which is a solution for (2.155). The iteration in (2.156), called the Peano-Picard-Lindelöf iteration, is actually one way of obtaining a solution for the differential equation (2.155). The uniqueness of the solution depends on the Lipschitz condition: e.g. $\dot{x} = \sqrt{x}$, $x(0) = 0$ does not satisfy the Lipschitz condition, and $\phi(t) = 0$, and $\phi(t) = t^2/4$ are two distinct solutions.

2.4.6 Linear homogeneous differential equations

The differential equation

$$\dot{\mathbf{x}} = \mathbf{A}(t)\mathbf{x}, \quad \mathbf{x}(t) \in \mathbb{R}^n, \quad \mathbf{x}(t_0) = \mathbf{x}_0, \quad \mathbf{A}(\cdot) \text{ piecewise continuous} \quad (2.157)$$

satisfies

1. For every fixed $\mathbf{x}$, $t \to \mathbf{A}(t)\mathbf{x}$ is piecewise continuous.

2. Lipschitz condition:

$$\|\mathbf{A}(t)\mathbf{x} - \mathbf{A}(t)\mathbf{x}'\| \leq \|\mathbf{A}(t)\|\|\mathbf{x} - \mathbf{x}'\| \quad \forall \mathbf{x}, \mathbf{x}' \in \mathbb{R}^n.$$

So, (2.157) has a unique solution. It can be shown that the set of solutions for (2.157), $\{\phi(t, t_0, \mathbf{x}_0) : \mathbf{x}_0 \in \mathbb{R}^n\}$, is an n-dimensional linear space. Let $\{\mathbf{x}_i\}_{i=1}^n$ be an arbitrary basis for $\mathbb{R}^n$. Define $\psi_i(t) = \phi(t, t_0, \mathbf{x}_i)$. Then, it can be shown that $\{\psi_i\}_{i=1}^n$ is a linearly independent set, and spans the set of solutions. For $\{\mathbf{x}_i\}_{i=1}^n$ linearly independent, define

$$\mathbf{X}(t, t_0) = [\phi(t, t_0, \mathbf{x}_1)|\ldots|\phi(t, t_0, \mathbf{x}_n)] \quad (2.158)$$

which is called a *fundamental matrix* for (2.157). If $\mathbf{x}_i = [0, \ldots, 0, 1, 0, \ldots, 0]$, with 1 at the i-th position, then the corresponding fundamental matrix is called the *state-transition matrix* for the differential equation (2.157), denoted as $\mathbf{\Phi}(t, t_0)$. It can be shown that

- $\mathbf{X}(t, t_0)$ is nonsingular for all t.

- For any $\mathbf{x}_0$, there exists a unique $\mathbf{c} \in \mathbb{R}^n$ such that $\phi(t, t_0, \mathbf{x}_0) = \mathbf{X}(t, t_0)\mathbf{c}$.

- Any two fundamental matrices are related as $\mathbf{X}_1(t, t_0) = \mathbf{X}_2(t, t_0)\mathbf{C}$ where $\mathbf{C}$ is a nonsingular $n \times n$ constant matrix.

- $\mathbf{\Phi}(t, t_0)$ is uniquely defined as the solution of $\dot{\mathbf{X}} = \mathbf{A}(t)\mathbf{X}$, $\mathbf{X}(t_0) = \mathbf{I}$

- The solution of (2.157) is given by $\phi(t, t_0, \mathbf{x}_0) = \mathbf{\Phi}(t, t_0)\mathbf{x}_0$ in terms of the state transition matrix and the initial condition.

2.4.7 Linear inhomogeneous differential equations

For the inhomogeneous linear differential equation

$$\dot{\mathbf{x}} = \mathbf{A}(t)\mathbf{x} + \mathbf{b}(t), \quad \mathbf{x}(t_0) = \mathbf{x}_0, \quad \mathbf{A}(.), \mathbf{b}(.) \text{ piecewise continuous} \quad (2.159)$$

the solution is given by

$$\phi(t, t_0, \mathbf{x}_0) = \mathbf{\Phi}(t, t_0)\mathbf{x}_0 + \int_{t_0}^{t} \mathbf{\Phi}(t, \tau)\mathbf{b}(\tau)d\tau \quad (2.160)$$

where $\mathbf{\Phi}(t, t_0)$ is the state transition matrix for the homogeneous equation (2.157).

2.4.8 Linear differential equations with constant coefficients

If $\mathbf{A}(t)$ in (2.157) is a constant matrix $\mathbf{A}(t) = \mathbf{A}$, i.e.

$$\dot{\mathbf{x}} = \mathbf{A}\mathbf{x}, \quad (2.161)$$

then the state transition matrix is given by

$$\mathbf{\Phi}(t, t_0) = \exp\left(\mathbf{A}(t - t_0)\right) = \mathbf{\Phi}(t - t_0) \quad (2.162)$$

Now, let us assume that the constant matrix $\mathbf{A}$ is diagonalizable (with distinct eigenvalues):

$$\mathbf{A} = \mathbf{P}\mathbf{D}\mathbf{P}^{-1} \quad (2.163)$$

where $\mathbf{D} = \text{diag.}(\lambda_i)$, and $\mathbf{P}$ is the modal matrix for $\mathbf{A}$:

$$\mathbf{P} = [\mathbf{x}_1, \mathbf{x}_2, \ldots, \mathbf{x}_n]. \quad (2.164)$$

Then

$$\begin{aligned}
\mathbf{\Phi}(t, 0) &= \exp\left(\mathbf{A}t\right) & (2.165) \\
&= \exp\left(\mathbf{P}\mathbf{D}\mathbf{P}^{-1}t\right) & (2.166) \\
&= \mathbf{P}\exp\left(\mathbf{D}t\right)\mathbf{P}^{-1} & (2.167)
\end{aligned}$$

and hence, $\mathbf{P}\exp(\mathbf{D}t)$ is a fundamental matrix for (2.161), which can be expressed as

$$\begin{aligned}
\mathbf{X}(t, 0) &= \mathbf{P}\exp(\mathbf{D}t) \\
&= [\exp(\lambda_1 t)\mathbf{x}_1, \exp(\lambda_2 t)\mathbf{x}_2, \ldots, \exp(\lambda_n t)\mathbf{x}_n].
\end{aligned}$$

We know that any solution of the differential equation (2.159) can be expressed as a linear combination of the columns of a fundamental matrix. So, any solution is in the form

$$\phi(t, 0, \mathbf{x}_0) = c_1 \exp(\lambda_1 t)\mathbf{x}_1 + c_2 \exp(\lambda_2 t)\mathbf{x}_2 + \ldots + c_n \exp(\lambda_n t)\mathbf{x}_n \quad (2.168)$$

where $c_1, \ldots, c_n \in \mathbb{R}$ depend on the initial condition, $\lambda_1, \ldots, \lambda_k$ are the eigenvalues of $\mathbf{A}$, and $\mathbf{x}_1, \mathbf{x}_2, \ldots, \mathbf{x}_n$ are the eigenvectors. This is called a *modal decomposition* of the solutions of (2.161), and each of $\exp(\lambda_i t)\mathbf{x}_i$ is called a *mode* of the system. For obvious reasons, the LTI system described by (2.161) is said to be *stable* if all of the eigenvalues of $\mathbf{A}$ have strictly negative real parts.

Now, let us concentrate on the differential equation

$$\dot{\mathbf{x}} = -\mathbf{A}^T \mathbf{x}. \quad (2.169)$$

Assuming that $\mathbf{A}$ is diagonalizable with distinct eigenvalues, and following analogous steps to the above ones, one can show that any solution of (2.169) is in the form

$$\psi(t, 0, \mathbf{x}_0) = c_1 \exp(-\lambda_1 t)\mathbf{y}_1 + c_2 \exp(-\lambda_2 t)\mathbf{y}_2 + \ldots + c_n \exp(-\lambda_n t)\mathbf{y}_n \quad (2.170)$$

where $c_1, \ldots, c_n \in \mathbb{R}$, $\lambda_1, \ldots, \lambda_k$ are the eigenvalues of $\mathbf{A}^T$ (and also $\mathbf{A}$), and $\mathbf{y}_1, \ldots, \mathbf{y}_n$ are the eigenvectors of $\mathbf{A}^T$. Recall that the eigenvectors of $\mathbf{A}$ and $\mathbf{A}^T$ satisfy (2.148). Furthermore, these eigenvectors can be chosen such that

$$\mathbf{y}_j^T \mathbf{x}_i = \begin{cases} 0 & \text{if } i \neq j \\ 1 & \text{if } i = j \end{cases} \quad (2.171)$$

Then, it is trivial to verify that the state transition matrix for the differential equation (2.161) can be expressed as

$$\Phi(t, \tau) = \Phi(t - \tau) = \sum_{i=1}^{n} \exp(\lambda_i (t - \tau)) \mathbf{x}_i \mathbf{y}_i^T. \quad (2.172)$$

The state transition matrix of (2.169), $\Psi(t - \tau)$, is then given by

$$\Psi(t - \tau) = \Phi(\tau - t)^T \quad (2.173)$$

in terms of the state transition matrix of (2.161).

2.4.9 Linear differential equations with periodic coefficients

Now, we consider the case when the coefficient matrix in (2.157) is periodically time-varying

$$\dot{\mathbf{x}} = \mathbf{A}(t)\mathbf{x}, \quad \mathbf{x}(t_0) = \mathbf{x}_0 \quad (2.174)$$
$$\mathbf{A}(t) = \mathbf{A}(t + T), \quad \text{for some period } T, \forall t.$$

Let $\Phi(t, t_0)$ be the state transition matrix for (2.174). It can be shown that it satisfies

1. $\Phi(t, t_0) = \Phi(t + nT, t_0 + nT) \quad \forall\, t, t_0$.

2. $\Phi(t + nT, t_0) = \Phi(t, t_0)\Phi(t_0 + nT, t_0) \quad \forall\, t, t_0$.

Let us define

$$\mathbf{F} = \frac{1}{T} \ln\left(\Phi(t_0 + T, t_0)\right), \quad \Phi(t_0 + T, t_0) = \exp\left(\mathbf{F}T\right). \qquad (2.175)$$

Now we present a result due to do G. Floquet (1883): Given (2.175), the state transition matrix of (2.174) can be expressed as

$$\Phi(t, t_0) = \mathbf{P}(t, t_0) \exp\left((t - t_0)\mathbf{F}\right) \qquad (2.176)$$

where $\mathbf{P}(.,t_0)$ is *periodic* with period T, i.e. $\mathbf{P}(t + T, t_0) = \mathbf{P}(t, t_0)$, and is *nonsingular*. Now, let us assume that the constant matrix $\mathbf{F}$ is diagonalizable (with distinct eigenvalues),

$$\mathbf{F} = \mathbf{UDU}^{-1} \qquad (2.177)$$

where $\mathbf{D} = \text{diag.}(\eta_i)$ with eigenvalues of $\mathbf{F}$ on the diagonal, and $\mathbf{U}$ is the modal matrix for $\mathbf{F}$ with the eigenvectors of $\mathbf{F}$ as the columns. The eigenvalues of $\mathbf{F}$ are called the *Floquet exponents* for (2.174). The eigenvalues of $\Phi(t_0 + T, t_0) = \exp\left(\mathbf{F}T\right)$ are given by

$$\mu_i = \exp\left(\eta_i T\right) \qquad (2.178)$$

in terms of the eigenvalues of $\mathbf{F}$. Then

$$\begin{aligned}
\Phi(t, t_0) &= \mathbf{P}(t, t_0) \exp\left((t - t_0)\mathbf{F}\right) & (2.179) \\
&= \mathbf{P}(t, t_0) \exp\left((t - t_0)\mathbf{UDU}^{-1}\right) & (2.180) \\
&= \mathbf{P}(t, t_0)\mathbf{U} \exp\left((t - t_0)\mathbf{D}\right)\mathbf{U}^{-1} & (2.181)
\end{aligned}$$

and hence $\mathbf{P}(t, t_0)\mathbf{U} \exp\left(\mathbf{D}(t - t_0)\right)$ is a fundamental matrix for (2.174), which can be expressed as

$$\begin{aligned}
\mathbf{X}(t, t_0) &= \mathbf{P}(t, t_0)\mathbf{U} \exp\left(\mathbf{D}(t - t_0)\right) & (2.182) \\
&= [\exp\left(\eta_1(t - t_0)\right)\mathbf{u}_1(t, t_0), \ldots, \exp\left(\eta_n(t - t_0)\right)\mathbf{u}_n(t, t_0)]
\end{aligned}$$

where $\mathbf{u}_i(t, t_0)$ are the columns of the matrix $\mathbf{P}(t, t_0)\mathbf{U}$, and hence, they are periodic in t with period T. We know that any solution of the differential equation can be expressed as a linear combination of the columns of a fundamental matrix. So, any solution is in the form

$$\phi(t, t_0, \mathbf{x}_0) = c_1 \exp\left(\eta_1(t - t_0)\right)\mathbf{u}_1(t, t_0) + \ldots + c_n \exp\left(\eta_n(t - t_0)\right)\mathbf{u}_n(t, t_0) \qquad (2.183)$$

where $c_1, \ldots, c_n \in \mathbb{R}$ depend on the initial condition. (2.183) is similar to (2.168) for the linear differential equations with constant coefficients. Note that
$$\mathbf{P}(t_0 + nT, t_0) = \mathbf{I}, \tag{2.184}$$
and hence $\mathbf{u}_i(t_0 + nT, t_0) = \bar{\mathbf{u}}_i$, where $\bar{\mathbf{u}}_i$ are the eigenvectors of $\mathbf{F}$. One important corollary is: (2.174) has a periodic solution with period T if and only if $\Phi(t_0 + T, t_0) = \exp(\mathbf{F}T)$ has an eigenvalue equal to 1. The LPTV system defined by (2.174) is said to be *stable* if all the Floquet exponents, i.e. the eigenvalues of $\mathbf{F}$, have strictly negative real parts. The eigenvalues of $\mathbf{F}$ for (2.174) play the same role as the eigenvalues of $\mathbf{A}$ do for (2.161) in determining the stability properties of the system. Note that, even if the eigenvalues of $\mathbf{A}(t)$ have strictly negative real parts for all t, the LPTV system described by (2.174) can be unstable. It is the eigenvalues of $\mathbf{F}$ that determine the stability properties of (2.174).

Analogous to the constant coefficient case, now we will look into the differential equation
$$\dot{\mathbf{x}} = -\mathbf{A}^T(t)\mathbf{x}. \tag{2.185}$$
With the definition of $\mathbf{F}$ in (2.175), it can be shown that the state transition matrix $\Psi(t, t_0)$ for (2.185) is given by
$$\Psi(t, t_0) = \mathbf{Q}(t, t_0) \exp\left(-(t - t_0)\mathbf{F}^T\right) \tag{2.186}$$
where $\mathbf{Q}(., t_0)$ is *periodic* with period T, i.e. $\mathbf{Q}(t + T, t_0) = \mathbf{Q}(t, t_0)$, and is *nonsingular*. Again assuming that $\mathbf{F}$ is diagonalizable with distinct eigenvalues, and following analogous steps to the ones for the untransposed equation system, one can show that any solution of (2.185) is in the form
$$\psi(t, t_0, \mathbf{x}_0) = c_1 \exp\left(-\eta_1(t - t_0)\right)\mathbf{v}_1(t, t_0) + \ldots + c_n \exp\left(-\eta_n(t - t_0)\right)\mathbf{v}_n(t, t_0) \tag{2.187}$$
where η_i are the eigenvalues of $\mathbf{F}$ (and hence $\mathbf{F}^T$), and $\mathbf{v}_i(t, t_0)$ are periodic in t with T, and satisfy $\mathbf{v}_i(t_0 + nT, t_0) = \bar{\mathbf{v}}_i$, where $\bar{\mathbf{v}}_i$ are the eigenvectors of $\mathbf{F}^T$ From (2.148), the eigenvectors $\bar{\mathbf{u}}_i$ and $\bar{\mathbf{v}}_i$ can be chosen to satisfy
$$\bar{\mathbf{v}}_j^T \bar{\mathbf{u}}_i = \begin{cases} 0 & \text{if } i \neq j \\ 1 & \text{if } i = j \end{cases} \tag{2.188}$$
We know that $\mathbf{w}_i(t) = \exp\left(\eta_i(t - t_0)\right)\mathbf{u}_i(t, t_0)$ is a solution of (2.174), hence satisfies $\dot{\mathbf{x}} = \mathbf{A}(t)\mathbf{x}$. Similarly, $\mathbf{z}_j(t) = \exp\left(-\eta_j(t - t_0)\right)\mathbf{v}_j(t, t_0)$ is a solution of (2.185), hence satisfies $\dot{\mathbf{x}} = -\mathbf{A}(t)^T\mathbf{x}$. Then, we have
$$\begin{aligned}
\frac{d}{dt}(\mathbf{z}_j(t)^T \mathbf{w}_i(t)) &= (\frac{d}{dt}\mathbf{z}_j(t)^T)\mathbf{w}_i(t) + \mathbf{z}_j(t)^T \frac{d}{dt}\mathbf{w}_i(t) & (2.189) \\
&= -\mathbf{z}_j(t)^T \mathbf{A}(t)\mathbf{w}_i(t) + \mathbf{z}_j(t)^T \mathbf{A}(t)\mathbf{w}_i(t) & (2.190) \\
&= 0. & (2.191)
\end{aligned}$$

From (2.188) we also know that

$$\mathbf{z}_j(t_0)^T \mathbf{w}_i(t_0) = \begin{cases} 0 & \text{if } i \neq j \\ 1 & \text{if } i = j \end{cases} \quad (2.192)$$

Combining (2.192) and (2.191), we get

$$\mathbf{z}_j(t)^T \mathbf{w}_i(t) = \exp\left((\eta_i - \eta_j)(t - t_0)\right) \mathbf{v}_j^T(t, t_0) \mathbf{u}_i(t, t_0) = \begin{cases} 0 & \text{if } i \neq j \\ 1 & \text{if } i = j \end{cases} \quad (2.193)$$

for all $t \geq t_0$, which implies

$$\mathbf{v}_i^T(t, t_0) \mathbf{u}_i(t, t_0) = \begin{cases} 0 & \text{if } i \neq j \\ 1 & \text{if } i = j \end{cases} \quad (2.194)$$

for all $t \geq t_0$.

Given all the above, one can easily verify that the state transition matrix for (2.174) can be expressed as [16]

$$\Phi(t, \tau) = \sum_{i=1}^{n} \exp\left(\eta_i(t - \tau)\right) \mathbf{u}_i(t) \mathbf{v}_i^T(\tau). \quad (2.195)$$

The state transition matrix of (2.185), $\Psi(t, \tau)$ is then given by

$$\Psi(t, \tau) = \Phi(\tau, t)^T \quad (2.196)$$

in terms of the state transition matrix of (2.174). Then,

$$\Phi(T, 0) = \sum_{i=1}^{n} \exp\left(\eta_i T\right) \mathbf{u}_i(T) \mathbf{v}_i^T(0) \quad (2.197)$$

$$= \sum_{i=1}^{n} \exp\left(\eta_i T\right) \mathbf{u}_i(0) \mathbf{v}_i^T(0) \quad (2.198)$$

since $\mathbf{u}_i(t)$ is periodic in t with period T. From (2.198), $\mathbf{u}_i(0)$ are the eigenvectors of $\Phi(T, 0)$ with corresponding eigenvalues $\exp(\eta_i T)$, and $\mathbf{v}_i(0)$ are the eigenvectors of $\Phi(T, 0)^T$ with corresponding eigenvalues $\exp(\eta_i T)$.

2.5 Stochastic Differential Equations and Systems[§]

In Section 2.3, we have presented a dynamical system representation from an *input/output* relationship perspective, considering the system as a *black box*,

---

[§]Most of the material in this section is summarized from [9] and [10], which are excellent books on the topic. Other useful references are [11], [12] and [13].

(impulse responses and transfer functions) to study the effect of the input noise signals modeled as stochastic processes on the system. However we did not discuss how one would arrive at these transfer functions from a physical description of the system. The systems we are going to deal with, i.e. electronics circuits and systems, are usually modeled with a system of ordinary differential equations. Other forms of system representation, e.g partial differential equations or finite-state machines, are also used for electronic circuits. For the type of circuits we are concerned with, i.e. analog and mixed-signal circuits, systems of differential equations, with some additions, are the most appropriate mathematical models for various levels of the design hierarchy. Since we model the noise signals as stochastic processes, a mathematical model of the system with a system of differential equations will involve stochastic processes. In this section, we will give an overview of the theory of stochastic differential equations and systems. The *transfer function* model of a system, and the *system of differential equations* model are closely related. One can convert a transfer function model to a system of differential equations model and vice versa, but this is not always practically feasible. Mathematical analysis of these models provide information about the system in different but related forms. Usually, using a combination of these two forms of mathematical models will provide the most useful information to the system designer. In a "loose" and nonrigorous sense, one can say that a system of differential equations model is more "precise" than a transfer function model. In engineering practice, transfer function models are widely used. Using a transfer function model (and the associated operational calculus) blindly can, sometimes, yield wrong results and obscure some properties of the system. We will give examples for this later. In principle, one should always exercise care in using the operational calculus associated with transfer function models, and make sure that the underlying differential equation model and its meaning are mathematically sound and rigorous.

2.5.1 Overview

Differential equations involving random variables and stochastic processes arise in the investigation of numerous physics and engineering problems. They are usually of one of the following two fundamentally different types. On the one hand, certain functions, coefficients, parameters, and boundary or initial values in classical differential equation problems can be random. Simple examples are

$$\dot{X}(t) = A(t)X(t) + B(t) \quad X(t_0) = C \quad (2.199)$$

with stochastic processes $A(t)$ and $B(t)$ as coefficients and with random initial value C, or

$$\dot{X}(t) = f(t, X(t), \zeta(t)) \quad X(t_0) = C \quad (2.200)$$

with the stochastic process $\zeta(t)$, the random initial value C, and the fixed deterministic function f. If these stochastic processes have certain regularity properties, one can consider the above-mentioned problems simply as a family of classical problems for the individual sample paths, and treat them with the classical methods of the theory of differential equations.

The situation is quite different if "stochastic processes" of the so-called "white noise" type appear in what is written formally as an ordinary differential equation, for example, the "stochastic process" $\xi(t)$ in the equation

$$\dot{X}(t) = f(t, X(t)) + G(t, X(t))\xi(t) \quad X(t_0) = C. \tag{2.201}$$

As discussed in Section 2.2.10, white noise $\xi(t)$ does not exist in the conventional sense, but it is widely used in engineering practice as a very useful mathematical idealization for describing random influences that fluctuate rapidly, and hence, are virtually uncorrelated for different instants of time.

Such equations were first treated in 1908 by Langevin in the study of the Brownian motion of a particle in a fluid. If $X(t)$ is a component of the velocity, at an instant t, of a free particle that performs a Brownian motion, Langevin's equation is

$$\dot{X}(t) = -\alpha X(t) + \sigma \xi(t) \quad \alpha > 0, \, \sigma \text{ constants}. \tag{2.202}$$

Here, $-\alpha X(t)$ is the systematic part of the influence of the surrounding medium due to dynamic friction. The constant α is found from Stoke's law to be $\alpha = 6\pi a \eta / m$, where a is the radius of the (spherical) particle, m is its mass, and η is the viscosity of the surrounding fluid. On the other hand, the term $\sigma \xi(t)$ represents the force exerted on the particle by the molecular collisions. Since, under normal conditions, the particle uniformly undergoes about 10^{21} molecular collisions per second from all directions, $\sigma \xi(t)$ is indeed a rapidly varying fluctuational term, which can be idealized as "white noise". The covariance for $\xi(t)$ is the Dirac's delta function, then $\sigma^2 = 2\alpha k T / m$ (where k is Boltzmann's constant and T is the absolute temperature of the surrounding fluid). Note that, (2.202) is a special case of (2.201), the right-hand member of which is decomposed as the sum of a systematic part f and a fluctuational $G\xi(t)$.

In the model of Brownian motion, (2.202), one can calculate explicitly the fdds of $X(t)$ even though $\xi(t)$ is not a stochastic process in the usual sense. Every process $X(t)$ with these distributions (Ornstein-Uhlenbeck process) is nondifferentiable in any useful sense (see Section 2.2.9), hence (2.202) and, more generally, (2.201) cannot be regarded as ordinary differential equations.

For a mathematically rigorous treatment of equations of type (2.201), a new theory was necessary. As we have seen it in Section 2.2.10, even though "white noise" is only a generalized stochastic process, the indefinite integral

$$W(t) = \int_0^t \xi(s)ds \qquad (2.203)$$

can be identified with the Wiener process, or any other zero-mean process with uncorrelated stationary increments. *Gaussian* white noise is identified as the formal derivative of the Wiener process.

If we write (2.203) symbolically as

$$dW(t) = \xi(t)dt \qquad (2.204)$$

(2.201) can be put in the differential form

$$dX(t) = f(t, X(t))dt + G(t, X(t))dW(t) \quad X(t_0) = C. \qquad (2.205)$$

This is a *stochastic differential equation* (SDE) for the process $X(t)$. It should be understood as an abbreviation for the integral equation

$$X(t) = C + \int_{t_0}^t f(s, X(s))ds + \int_{t_0}^t G(s, X(s))dW(s). \qquad (2.206)$$

The second integral in (2.206) cannot, even for smooth G, be regarded in general as an ordinary Riemann-Stieltjes integral, because the value depends on the intermediate points in the approximating sums. We will elaborate on this later. In 1951, K. Ito defined integrals of the form

$$Y(t) = \int_{t_0}^t G(s)dW(s) \qquad (2.207)$$

for a broad class of so-called "nonanticipating" functionals of the Wiener process $W(t)$, and in doing so, put the theory of stochastic differential equations on a solid foundation. This theory has its peculiarities. For example, the solution of the stochastic differential equation

$$dX(t) = X(t)dW(t) \quad X(0) = 1 \qquad (2.208)$$

is not $\exp(W(t))$, but

$$X(t) = \exp(W(t) - t/2) \qquad (2.209)$$

which can not be derived by formal calculation according to the classical rules of calculus.

It turns out that the solution of a stochastic differential equation of the form (2.205) is a Markov process with continuous sample paths, which is called a *diffusion process*. In a loose sense, diffusion processes are "smooth" Markov processes. The Markov property discussed in Section 2.2.3 is basically the *causality* principle of classical physics carried over to stochastic dynamic systems. It specifies that the knowledge of the state of a system at a given time is sufficient to determine its state at any future time. For instance, the ordinary differential equation

$$\dot{x} = f(t, x) \tag{2.210}$$

states that the change taking place in $x(t)$ at time t depends only on $x(t)$ and t, and not on the values of $x(s)$ for $s < t$. A consequence of this is that, under certain conditions on f, discussed in Section 2.4.5, the solution curve for $x(t)$ is uniquely determined by an initial point (t_0, c)

$$x(t) = x(t, t_0, c), \quad t > t_0, \quad x(t_0) = c. \tag{2.211}$$

If this idea is carried over to stochastic dynamic systems, we get the Markov property. It says that if the state of the system at a particular time s (the present) is known, additional information regarding the behavior of the system at times $t < s$ (the past) has no effect on our knowledge of the *probable* development of the system at $t > s$ (in the future).

For Markov and diffusion processes, there exist effective methods for calculating transitional probabilities and fdds, which deal with the timewise development of transition probabilities $P(X(t) \in \mathcal{B}|X(s) = x)$, where $\mathcal{B} \subset \mathbb{R}$. In contrast, the calculus of stochastic differential equations deals with the random variable $X(t)$ and its variation. An equation of the form (2.205) or (2.206) represents a construction rule with which one can construct the trajectories of $X(t)$ from the trajectories of a Wiener process $W(t)$ and an initial value C.

The law of motion for the state of a Markovian stochastic dynamic system can be described by an equation of the form

$$dX(t) = g(t, X(t), dt). \tag{2.212}$$

In the case of fluctuational influences, i.e noise, that are additively superimposed on a systematic part (which will be the case for noise in electronic circuits), we have

$$g(t, x, h) = f(t, x)h + G(t, x)(Y(t + h) - Y(t)). \tag{2.213}$$

Here, $Y(t)$ is a process with uncorrelated stationary increments and (2.212) takes the form

$$dX(t) = f(t, X(t))dt + G(t, X(t))dY(t). \tag{2.214}$$

$Y(t)$ is any process with uncorrelated stationary increments. In particular, $Y(t)$ could be the Wiener process, or the Poisson process (with mean subtracted). Due to the central limit theorem, our models of white noise will be assumed to be Gaussian, which is then the symbolic derivative of the Wiener process.

2.5.2 An example

Now, we will analyze the following differential equation to illustrate how using classical calculus can yield wrong results when a "white noise" type driving source is present[2]:

$$\frac{d}{dt}I(t) = -\alpha I(t) + q\frac{d}{dt}N(t) \qquad (2.215)$$

where $N(t)$ is a homogeneous Poisson process with parameter λ (see Section 2.2.8). This equation is actually a model for shot noise in electronic circuit devices. This is a kind of stochastic differential equation similar to Langevin's equation (2.202), in which, however, the fluctuational force is given by $\mu(t) = qd/dt N(t)$. The mean of $\mu(t)$ is nonzero, in fact

$$\mathbf{E}\left[\mu(t)dt\right] = \mathbf{E}\left[dN(t)\right] = \lambda dt, \qquad (2.216)$$
$$\mathbf{E}\left[(dN(t) - \lambda dt)^2\right] = \lambda dt \qquad (2.217)$$

from the properties of the Poisson distribution (variance equals the mean) and the Poisson process discussed before. We, then, define the fluctuation as the difference between the mean value and $dN(t)$

$$d\nu(t) = dN(t) - \lambda dt, \qquad (2.218)$$

so the stochastic differential equation (2.215) takes the form

$$dI(t) = (\lambda q - \alpha I(t))dt + q d\nu(t). \qquad (2.219)$$

Now, let us use ordinary calculus to solve this equation. If we take the expectation of both sides of (2.219) and exchange the order of differentiation and expectation, we get

$$\frac{d}{dt}\mathbf{E}\left[I(t)\right] = \lambda q - \alpha \mathbf{E}\left[I(t)\right], \qquad (2.220)$$

since $\nu(t)$ is a zero-mean process. Now, we will use the following differentiation rule of ordinary calculus:

$$d(I^2) = 2I dI. \qquad (2.221)$$

Using (2.221),

$$\begin{aligned}\frac{1}{2}d(I(t)^2) &= I(t)dI(t) & (2.222)\\ &= I(t)((\lambda q - \alpha I(t))dt + q d\nu(t)) & (2.223)\\ &= (\lambda q I(t) - \alpha I(t)^2)dt + q I(t) d\nu(t), & (2.224)\end{aligned}$$

where we have used (2.219) for $dI(t)$. If we take the expectation of both sides of (2.224) we get

$$\frac{1}{2}\frac{d}{dt}\mathrm{E}\left[I(t)^2\right] = \lambda q \mathrm{E}\left[I(t)\right] - \alpha \mathrm{E}\left[I(t)^2\right], \qquad (2.225)$$

where we have used $\mathrm{E}\left[I(t)d\nu(t)\right] = 0$. Now, we will evaluate (2.220) and (2.225) at *steady-state*, that is in the limit $t \to \infty$. Setting $\frac{d}{dt}\mathrm{E}\left[I(t)\right] = 0$ in (2.220) we get

$$\mathrm{E}\left[I(\infty)\right] = \lambda q / \alpha, \qquad (2.226)$$

which is reasonable, it gives us the average current through the system. Then we set $\frac{d}{dt}\mathrm{E}\left[I(t)^2\right] = 0$, and use (2.226) in (2.225) to get

$$\mathrm{E}\left[I(\infty)^2\right] = (\lambda q / \alpha)^2, \qquad (2.227)$$

which says that the mean-square current is the same as the square of the mean, i.e. the variance of the current at $t \to \infty$ is zero. This is a rather unreasonable result. Now, we reexamine the differentiation rule (2.221) we used. In deriving (2.221), one writes

$$d(I^2) = (I + dI)^2 - I^2 = 2IdI + (dI)^2 \qquad (2.228)$$

and then drops the term $(dI)^2$ as being of second order in dI. However, now recall (2.217) and (2.218), which say that

$$\mathrm{E}\left[(d\nu(t))^2\right] = \lambda dt, \qquad (2.229)$$

so that a quantity of *second order* in $d\nu$ is actually of first order in dt. A sample path of $N(t)$ is a step function, discontinuous, and certainly not differentiable at the times of the arrivals. In the ordinary sense, none of these calculus manipulations are permissible. Now, let us rederive the differentiation rule (2.221) by paying attention to (2.229):

$$\begin{aligned}\mathrm{E}\left[d(I^2)\right] &= \mathrm{E}\left[(I+dI)^2 - I^2\right] = \mathrm{E}\left[2IdI + (dI)^2\right] \\ &= 2\mathrm{E}\left[I((\lambda q - \alpha I(t))dt + qd\nu(t))\right] \\ &\quad + \mathrm{E}\left[((\lambda q - \alpha I(t))dt + qd\nu(t))^2\right].\end{aligned} \qquad (2.230)$$

Again using $\mathrm{E}\left[I(t)d\nu(t)\right] = 0$, with (2.229), and dropping terms that are higher than first order in dt, we obtain

$$\frac{1}{2}d\mathrm{E}\left[I(t)^2\right] = (\lambda q \mathrm{E}\left[I(t)\right] - \alpha \mathrm{E}\left[I(t)^2\right] + \frac{q^2\lambda}{2})dt, \qquad (2.231)$$

and as $t \to \infty$, this gives

$$\mathrm{E}\left[I(\infty)^2\right] - (\mathrm{E}\left[I(\infty)\right])^2 = \frac{q^2\lambda}{2\alpha}. \tag{2.232}$$

So, the variance of $I(t)$ does not go to zero as $t \to \infty$. The conclusion is that stochastic processes can not normally be differentiated according to the usual laws of calculus. Special rules have to be developed when dealing with "white noise" type of processes, and a precise specification of what one means by differentiation becomes important.

2.5.3 Stochastic integrals

The analysis of stochastic systems often leads to differential equations of the form

$$\dot{X}(t) = f(t, X(t)) + G(t, X(t))\xi(t), \quad X(t_0) = C, \tag{2.233}$$

where $\xi(t)$ is a "white noise" type of process. From now on, we will assume that $\xi(t)$ is Gaussian, i.e. the formal derivative of the Wiener process. For notational simplicity, we will concentrate on scalar equations of the form (2.233). All of the results we are going to present can be translated to the multidimensional case in a straightforward manner. As we discussed, $\xi(t)$ does not exist in a strict mathematical sense, but it is interpreted as the derivative of a well-defined process, i.e. the Wiener process.

The solution of a deterministic differential equation

$$\dot{x}(t) = f(t, x(t)), \quad x(t_0) = c \tag{2.234}$$

(with some conditions on f) is equivalent to the solution of the integral equation

$$x(t) = c + \int_{t_0}^{t} f(s, x(s))ds \tag{2.235}$$

for which it is possible to find a solution by means of the classical iteration procedure. In the same way, (2.233) is transformed into an integral equation

$$X(t) = C + \int_{t_0}^{t} f(s, X(s))ds + \int_{t_0}^{t} G(s, X(s))\xi(s)ds. \tag{2.236}$$

Then, $\xi(s)ds$ above is replaced by $dW(s)$ where $W(s)$ is the Wiener process to obtain

$$X(t) = C + \int_{t_0}^{t} f(s, X(s))ds + \int_{t_0}^{t} G(s, X(s))dW(s). \tag{2.237}$$

Now, (2.237) involves only well-defined stochastic processes. The second-integral above can not, in general, be interpreted as an ordinary Riemann-Stieltjes integral. Now, we move on to discuss the integrals of the type

$$\int_{t_0}^{t} G(s)dW(s) \qquad (2.238)$$

where $G(t)$ is an arbitrary stochastic process. The integral above is defined as a kind of Riemann-Stieltjes integral. Namely, the interval $[t_0, t]$ is divided into n subintervals by means of partitioning points

$$t_0 < t_1 < \cdots < t_{n-1} < t, \qquad (2.239)$$

and the intermediate points τ_i are defined such that

$$t_{i-1} \leq \tau_i \leq t_i. \qquad (2.240)$$

The *stochastic integral* $\int_{t_0}^{t} G(s)dW(s)$ is defined as the *limit* of the partial sums

$$S_n = \sum_{i=1}^{n} G(\tau_i)(W(t_i) - W(t_{i-1})) \qquad (2.241)$$

where the limit is to be taken using one of the limit notions discussed in Section 2.1.4. In general, the integral defined as the limit of S_n depends on *the particular choice of the intermediate points τ_i*. For instance, if $G(t) = W(t)$ then, using the properties of the Wiener process, it can be shown that

$$\mathbf{E}\left[S_n\right] = \sum_{i=1}^{n} (\tau_i - t_{i-1}) \qquad (2.242)$$

and

$$S_n \xrightarrow{2} (W(t)^2 - W(t_0)^2)/2 - (t - t_0)/2 + \sum_{i=1}^{n} (\tau_i - t_{i-1}) \quad \text{as } n \to \infty \qquad (2.243)$$

where we used the notion of *mean-square* limit. (See Section 2.1.4 for the definition of mean-square limit that is denoted by "$\xrightarrow{2}$".) Therefore, in order to obtain a *unique* definition of the stochastic integral, it is necessary to define specific intermediate points τ_i. For example, if we choose

$$\tau_i = \alpha t_i + (1 - \alpha)t_{i-1}, \quad 0 \leq \alpha \leq 1, \quad i = 1, 2, \ldots, n \qquad (2.244)$$

then

$$S_n \xrightarrow{2} (W(t)^2 - W(t_0)^2)/2 + (\alpha - 1/2)(t - t_0) \quad \text{as } n \to \infty \qquad (2.245)$$

and the stochastic integral is defined as this limit which depends on the choice of α:

$$((\alpha)) \int_{t_0}^{t} W(s)dW(s) = (W(t)^2 - W(t_0)^2)/2 + (\alpha - 1/2)(t - t_0). \quad (2.246)$$

"$((\alpha))$" in front of the integral in (2.246) denotes that the integral definition depends on the choice of α. For general $G(t)$, the *stochastic integral* is defined as the mean-square limit

$$S_n = \sum_{i=1}^{n} G(\tau_i)(W(t_i) - W(t_{i-1})) \xrightarrow{2} ((\alpha)) \int_{t_0}^{t} G(s)dW(s) \quad \text{as } n \to \infty. \quad (2.247)$$

For a choice of $\alpha = 0$, that is $\tau_i = t_{i-1}$, the stochastic integral defined above is called the *Ito stochastic integral*. For the rest of our discussion, unless otherwise stated, we will be dealing with Ito stochastic integrals. For $G(t) = W(t)$, with the Ito stochastic integral, we obtain

$$(Ito) \int_{t_0}^{t} W(s)dW(s) = (W(t)^2 - W(t_0)^2)/2 - (t - t_0)/2 \quad (2.248)$$

which does not agree with

$$\int_{t_0}^{t} W(s)dW(s) = (W(t)^2 - W(t_0)^2)/2 \quad (2.249)$$

which we would get using the ordinary Riemann-Stieltjes integral. The reason for this is that $|W(t + \triangle t) - W(t)|$ is almost always of the order of $\sqrt{\triangle t}$, so that in contrast to ordinary integration, terms of second order in $\triangle W(t)$ do not vanish on taking the limit.

It is disconcerting that the result (2.248) obtained by the Ito stochastic integral does not coincide with (2.249) obtained by formal application of the classical rules. The applicability of the rules of classical Riemann-Stieltjes calculus with a choice of $\alpha = 1/2$ in (2.247) (and hence in (2.246)) was the motivation for the definition of a stochastic integral given by R. L. Stratonovich. Again for $G(t) = W(t)$, Stratonovich's definition (with $\alpha = 1/2$) yields

$$(Strat) \int_{t_0}^{t} W(s)dW(s) = (W(t)^2 - W(t_0)^2)/2. \quad (2.250)$$

We will return to the discussion of Stratonovich's stochastic integral and its comparison to the Ito stochastic integral in the next section. Now, we concentrate on the Ito stochastic integral.

The stochastic process $G(t)$ in the definition of the stochastic integral (2.247) is called *nonanticipating*, if for all s and t such that $t < s$, the random variable $G(t)$ is probabilistically independent of $W(s) - W(t)$. This means that $G(t)$ is independent of the behavior of the Wiener process in the future of t. Since we are studying differential equations involving time which are supposed to model real physical systems that are causal (in the sense that the unknown future can not affect the present), we will restrict ourselves to nonanticipating $G(t)$ in the definition of the stochastic integral. As we shall see in the next section, this is required to define stochastic differential equations.

One can show that the Ito stochastic integral $\int_{t_0}^{t} G(s)dW(s)$ exists whenever the stochastic process $G(t)$ is nonanticipating and satisfies some "smoothness" conditions on the closed interval $[t_0, t]$.

It can also be shown that the Ito stochastic integral satisfies the following formula

$$\int_{t_0}^{t} G(s)(dW(s))^{2+N} = \begin{cases} \int_{t_0}^{t} G(s)ds & \text{for } N = 0 \\ 0 & \text{for } N > 0 \end{cases} \quad (2.251)$$

for an arbitrary nonanticipating stochastic process $G(t)$, which means that

$$\sum_{i=1}^{n} G(t_{i-1})(W(t_i) - W(t_{i-1}))^{2+N} \xrightarrow{2} \begin{cases} \int_{t_0}^{t} G(s)ds & \text{for } N = 0 \\ 0 & \text{for } N > 0 \end{cases} \quad \text{as } n \to \infty. \quad (2.252)$$

Similarly, one can show that

$$\int_{t_0}^{t} G(s)ds dW(s) = 0 \quad (2.253)$$

in the mean-square limit sense. The simplest way of characterizing these results is to say that $dW(t)$ is an infinitesimal order of $1/2$ and that in calculating differentials, infinitesimals of order higher than 1 are discarded.

2.5.4 Stochastic differential equations

We will concentrate on a scalar stochastic differential of the form

$$dX(t) = f(t, X(t)) + G(t, X(t))dW(t), \quad X(t_0) = C, \quad t_0 \leq t \leq T < \infty, \quad (2.254)$$

or in the integral form

$$X(t) = C + \int_{t_0}^{t} f(s, X(s))ds + \int_{t_0}^{t} G(s, X(s))dW(s), \quad t_0 \leq t \leq T < \infty, \quad (2.255)$$

where $X(t)$ is a stochastic process (assumed unknown for now) defined on $[t_0, T]$ and $W(t)$ is the Wiener process. The functions $f(t, x)$ and $G(t, x)$ are assumed to be deterministic for fixed t and x, i.e. the "randomness" in $f(t, X(t))$ and $G(t, X(t))$ appears only indirectly through $X(t)$. The fixed functions f and G determine the "system", and we have the two independent random elements: The initial condition C, and the Wiener process $W(t)$ modeling (as its formal derivative) a white "noise source" in the "system". We will interpret the second integral in (2.255) as the Ito stochastic integral, and later comment on the Stratonovich interpretation. Then, (2.255) is called an (Ito's) *stochastic differential equation*. Now, we consider conditions that need to be satisfied for the existence of a unique solution. An ordinary differential equation is a special case of (2.255) with $G = 0$ and C deterministic. The sufficient conditions for the existence of a unique solution are similar to the ones we have presented for an ordinary differential equation in Section 2.4.5. Suppose the functions $f(t, x)$ and $G(t, x)$ are defined on $[t_0, T] \times \mathbb{R}$ and have the following properties: There exists a constant $K > 0$ such that

1. (Lipschitz condition) for all $t \in [t_0, T]$, $x, y \in \mathbb{R}$

$$|f(t, x) - f(t, y)| + |G(t, x) - G(t, y)| \leq K|x - y|. \tag{2.256}$$

2. (Restriction on growth) for all $t \in [t_0, T]$, $x \in \mathbb{R}$

$$|f(t, x)|^2 + |G(t, x)|^2 \leq K^2(1 + |x|^2). \tag{2.257}$$

Then, (2.255) has on $[t_0, T]$ a unique solution $X(t)$, which is a stochastic process that is *almost surely continuous* (definition is analogous to mean-square continuity), and it satisfies the initial condition $X(t_0) = C$. If $X(t)$ and $Y(t)$ are almost-surely continuous solutions with the same initial value C, then

$$P\left(\sup_{t_0 \leq t \leq T} |X(t) - Y(t)| > 0\right) = 0. \tag{2.258}$$

Existence of the solution is proven using an analogous iteration to the Peano-Picard-Lindelöf iteration that is used for an ordinary differential equation as discussed in Section 2.4.5. We have already discussed the importance of the Lipschitz condition in Section 2.4.5. Failing to satisfy the Lipschitz condition can cause (2.255) to have distinct (i.e. nonunique) solutions. If the restricted growth condition is violated, the solution may "explode" in a finite time in the time interval $[t_0, T]$, which is a *random* finite time for (2.255). For instance, the following ordinary differential equation (which does not satisfy the restricted growth condition)

$$\dot{x} = x^2, \quad x(0) = c \tag{2.259}$$

has the solution
$$x(t) = (1/c - t)^{-1} \quad (2.260)$$
for $c \neq 0$. For $c > 0$, the solution explodes to ∞ at $t = 1/c$.

Now, let us consider an arbitrary function of $X(t)$: $h(X(t))$. We would like to find out what stochastic differential equation $h(X(t))$ obeys. Now, we will use (2.251) and (2.253) to expand $dh(X(t))$ to second order in $dW(t)$:

$$\begin{align}
dh(X(t)) &= h(X(t) + dX(t)) - h(X(t)) \quad (2.261)\\
&= h'(X(t))dX(t) + 1/2 h''(X(t))(dX(t))^2 + \ldots \quad (2.262)\\
&= h'(X(t))[f(t, X(t)) + G(t, X(t))dW(t)] \quad (2.263)\\
&\quad + 1/2 h''(X(t)) G^2(t, X(t))(dW(t))^2 + \ldots
\end{align}$$

where all other terms have been discarded since they are of higher order. Now we use $dW(t)^2 = dt$ to obtain

$$\begin{align}
dh(X(t)) = &[f(t,X(t))h'(X(t)) + 1/2 G^2(t,X(t))h''(X(t))]dt + \quad (2.264)\\
&G(t,X(t))h'(X(t))dW(t)
\end{align}$$

This formula is a special case of a general formula on stochastic differentials known as *Ito's formula*, and shows that the rule for changing variables is not given by ordinary calculus unless $h(x)$ is linear in x. The general form of Ito's formula for the multivariate case is quite complicated, and the easiest method is to simply use the multivariate form of the rule that $dW(t)$ is an infinitesimal of order $1/2$. It can be shown that for an n dimensional Wiener process $\mathbf{W}(t)$

$$\begin{align}
(d\mathbf{W}(t))d\mathbf{W}(t)^T &= \mathbf{I}_n dt, \quad (2.265)\\
dt\, d\mathbf{W}(t) &= 0 \quad (2.266)
\end{align}$$

which is obtained assuming that the components of the multidimensional Wiener process $\mathbf{W}(t)$ are independent.

It can be shown that the solution $X(t)$ of (2.254) is a Markov processes. The keystone for the Markov property of the solution is the fact that the "white noise" $\xi(t) = \frac{d}{dt} W(t)$ is a "process" with independent values at every time point.

2.5.5 Ito vs. Stratonovich

The Ito stochastic integral is mathematically and technically the most satisfactory, but, unfortunately, it is not always the most natural choice physically. The Stratonovich integral is the natural choice for an interpretation which assumes $\xi(t)$ is a real noise (not "white" noise) with finite correlation time, which is then allowed to become infinitesimally small after calculating measurable quantities. Hence, the system-theoretic significance of Stratonovich equations consists in the fact that, in many cases, they present themselves automatically when one approximates a white noise or a Wiener process with smoother processes, solves the approximating equation, and in the solution shifts back to white noise. Furthermore, a Stratonovich interpretation enables us to use ordinary calculus, which is not possible for an Ito interpretation. On the other hand, from a mathematical point of view, the choice is made clear by the near impossibility of carrying out proofs using the Stratonovich integral.

It is true that for the same stochastic differential equation (2.254), one can obtain different solutions using the Ito and Stratonovich interpretations of the stochastic integral. This discrepancy arises *not* from errors in the mathematical calculation but from a general discontinuity of the relationship between differential equations for stochastic processes and their solutions.

If the second integral in (2.255) is interpreted using the Stratonovich stochastic integral, then (2.254) is called a Stratonovich stochastic differential equation. It can be shown that the Ito stochastic differential equation

$$dX = f dt + G dW(t) \tag{2.267}$$

is the same as the Stratonovich stochastic differential equation

$$dX = (f - 1/2 G \partial_x G) dt + G dW(t) \tag{2.268}$$

or conversely, the Stratonovich stochastic differential equation

$$dX = \bar{f} dt + \bar{G} dW(t) \tag{2.269}$$

is the same as the Ito stochastic differential equation

$$dX = (\bar{f} + \frac{1}{2} \bar{G} \partial_x \bar{G}) dt + \bar{G} dW(t). \tag{2.270}$$

Thus, whether we think of a given formal equation (2.254) in the sense of Ito or Stratonovich, we arrive at the same solution as long as $G(t,x) = G(t)$ is independent of x. In general, we obtain two distinct Markov processes as solutions, which differ in the systematic (drift) behavior but not in the fluctuational (diffusion) behavior. If we consider $f(t,x)$ to be the model of the "systematic", or "large-signal" behavior of the system, and $G(t,x)$ as the effect of the state of the system on the intensity of a noise source modeled by the derivative of the Wiener process, then we can say that switching between the Ito and Stratonovich interpretations of the stochastic differential equation (2.254) is equivalent to changing the model of the systematic/deterministic behavior of the system on the "order" of the noise source intensity, and keeping the noise source intensity the same. For most of the practical physical systems, the noise signals are small compared with the deterministic/desired signals in the system. Hence, from a practical point of view, the choice between the Ito and Stratonovich interpretations of a stochastic differential equation is not a significant issue as it is from a technical point of view. For the rest of our discussion, we will use the Ito interpretation of the stochastic integral and Ito calculus for its nice mathematical properties.

2.5.6 Fokker-Planck equation

Consider the time development of an arbitrary $h(X(t))$, where $X(t)$ satisfies the stochastic differential equation (2.254). Using Ito's formula in (2.264), we obtain

$$\mathrm{E}\left[dh(X(t))\right]/dt = \mathrm{E}\left[\frac{d}{dt}h(X(t))\right] = \frac{d}{dt}\mathrm{E}\left[h(X(t))\right] \qquad (2.271)$$

$$= \mathrm{E}\left[f(t, X(t))\partial_x h + \frac{1}{2}G^2(t, X(t))\partial_x^2 h\right] \qquad (2.272)$$

Now, assume that $X(t)$ has a *conditional* probability density $p(x, t|x_0, t_0)$. Then

$$\begin{array}{rl} \frac{d}{dt}\mathrm{E}\left[h(X(t))\right] &= \int h(x)\partial_t p(x, t|x_0, t_0)dx \\ &= \int [f(t, x)\partial_x h + \frac{1}{2}G^2(t, x)\partial_x^2 h]p(x, t|x_0, t_0)dx. \end{array} \qquad (2.273)$$

One can proceed to obtain

$$\int h(x)\partial_t p\, dx = \int h(x)[-\partial_x(f(t, x)p) + 1/2\partial_x^2(G^2(t, x)p)]dx \qquad (2.274)$$

and, since $h(x)$ is arbitrary, we obtain

$$\partial_t p(x, t|x_0, t_0) = -\partial_x[f(t, x)p(x, t|x_0, t_0)] + \frac{1}{2}\partial_x^2[G^2(t, x)p(x, t|x_0, t_0)]. \qquad (2.275)$$

This is a partial differential equation for the conditional probability density $p(x, t|x_0, t_0)$ of the state $X(t)$ of the solution of (2.254). This equation is called the *Fokker-Planck equation*, or the *forward Kolmogorov equation* in one dimension. It can easily be generalized to the multivariate case. The conditional probability density as the solution of (2.275) is said to be the conditional probability density of a *diffusion process* with *drift* and *diffusion* coefficients given by $f(t, x)$ and $G^2(t, x)$ respectively. One can "loosely" define a diffusion process to be a Markov process with continuous sample paths. Then the drift $\mu(x, t)$ and the diffusion $\sigma^2(x, t)$ coefficients of a stochastic process X are defined by

$$\mathrm{E}\left[X(t+\epsilon) - X(t)|X(t) = x\right] = \mu(t, x) + o(\epsilon), \qquad (2.276)$$
$$\mathrm{E}\left[|X(t+\epsilon) - X(t)|^2|X(t) = x\right] = \sigma^2(t, x) + o(\epsilon). \qquad (2.277)$$

Wiener process $W(t)$ is a diffusion process with drift and diffusion coefficients given by $\mu = 0$ and $\sigma = 1$. If $X(t)$ is the solution of an (Ito) stochastic differential equation (2.254), then it is a diffusion process with drift and diffusion coefficients $f(t,x)$ and $G^2(t,x)$ respectively.

2.5.7 Numerical solution of stochastic differential equations

Effective analytical solutions of stochastic differential equations are most of the time achievable only in some simple cases. So, there is great interest in extending numerical integration techniques that are widely used for ordinary deterministic differential equations for use with stochastic differential equations. In this context, numerical simulation of a stochastic differential equation is meant to mean generating sample paths of the solution stochastic process $X(t)$ for the stochastic differential equation (2.254) at some discrete time points in a specified time interval. We will refer to this technique as the *direct* numerical integration of a stochastic differential equation to differentiate it from other numerical methods we will discuss later that are associated with stochastic differential equations. We would like to emphasize the fact that even though the *stochastic* numerical integration schemes are based on the ones that are used for ordinary differential equations, the extension to the stochastic case is by no means trivial. One has to exercise great care in the direct numerical integration of stochastic differential equations, and avoid using ad-hoc and unjustified extensions of deterministic numerical integration schemes. For instance, in contrast to the deterministic case, where different numerical methods converge (if they are convergent) to the same solution, in the case of the stochastic differential equation (2.254), different schemes can converge to *different* solutions (for the same noise source sample path and the initial condition). One also needs to consider various notions of convergence.

The final goal of numerically generating sample paths of the solution $X(t)$ for (2.254) at discrete time points is usually achieved in three steps. In the first step, the task is to design a numerical integration scheme so that the approximate solution generated at discrete time points converges to the real solution in some useful sense. As an example, let us consider the scalar Ito stochastic differential equation (2.254). Let the discretization of the interval $[t_0, T]$ be

$$t_0 < t_1 < \cdots < t_n = T, \tag{2.278}$$

and define the time increment as $\Delta t_i = t_{i+1} - t_i = h_i$, the increment of the standard Wiener process as $\Delta W_i = W_{i+1} - W_i$, and the approximation process as $\bar{X}(t_i) = \bar{X}_i$. Now let us consider the integral version of (2.254) on the interval $[t_i, t_{i+1}]$

$$X(t_{i+1}) = X(t_i) + \int_{t_i}^{t_{i+1}} f(s, X(s))ds + \int_{t_i}^{t_{i+1}} G(s, X(s))dW(s). \qquad (2.279)$$

Approximate schemes are constructed in such a way to approximate the integrals in (2.279) in some useful sense. For instance, for an approximation in the mean-square sense, for each $t \in [t_0, T]$ we would like to have

$$\mathsf{E}\left[|X(t) - \bar{X}(t)|^2\right] \to 0 \quad \text{as} \quad \max(h_i) \to 0 \qquad (2.280)$$

to assure convergence. Moreover, for fixed small $h = h_i$ (i.e. for large n), we would like to have $\mathsf{E}\left[|X(T) - \bar{X}_n|^2\right]$ to be small in some useful sense, e.g. we would like to have a numerical scheme such that, for all $t \in [t_0, t]$

$$\mathsf{E}\left[|X(t) - \bar{X}(t)|^2\right] = O(h^r) \qquad (2.281)$$

for $r \geq 1$. The simplest numerical scheme for the Ito stochastic differential equation (2.254) is the stochastic analog of the forward Euler scheme, which is

$$\bar{X}_{i+1} = \bar{X}_i + f(t_i, \bar{X}_i)h + G(t_i, \bar{X}_i) \triangle W_i. \qquad (2.282)$$

It can be shown that the stochastic forward Euler scheme converges in the mean-square sense to $X(t)$ governed by (2.254) as $h \to 0$. It can also be shown that, the order of convergence of the forward Euler scheme is $O(h)$, that is, for each $t \in [t_0, T]$

$$\mathsf{E}\left[|X(t) - \bar{X}(t)|^2\right] = O(h). \qquad (2.283)$$

The stochastic forward Euler scheme is not suitable for most practical problems. We will not further discuss other stochastic numerical integration schemes here. We would like to point out that another essential feature of numerical integration schemes for stochastic differential equations manifests itself in additional methodical difficulties when one wishes to deal with multidimensional equations. Again, one has to exercise great care in extending deterministic numerical schemes to the stochastic case.

The second step, for achieving the final goal of numerically generating a time discretized approximation of the sample paths of the solution stochastic process, is to replace the Wiener process with a suitable approximate simulation. There are various ways to accomplish this task. One way is to use a sequence of so-called *transport* processes. Let $Z_n(t)$, $n = 1, 2, \ldots$ be a sequence of continuous, piecewise linear "functions" with alternating slopes n and $-n$ with $Z_n(0) = 0$. The times between consecutive slope changes are *independent, exponentially distributed* random variables with parameter n^2. It can be shown that the transport process $Z_n(t)$ converges to the Wiener process $W(t)$, in a useful sense, as $n \to \infty$ in finite time intervals. It can also be shown that when $W(t)$ in (2.254) is replaced by $Z_n(t)$, the solution of the stochastic differential equation (2.254) converges to the real solution in a useful sense, provided that $f(t,x)$ and $G(t,x)$ in (2.254) satisfy certain conditions.

In the third step, one combines a stochastic numerical integration scheme with a method to simulate the noise source, i.e. the Wiener process. This can be interpreted as mixing two limiting procedures, one of which approximates the Wiener process and the other approximates the stochastic integration. This might lead to a result which may *not* coincide with the actual solution of the stochastic differential equation in a useful limiting sense. Special care has to be taken to make sure that the sample paths generated by the overall numerical algorithm converge to the sample paths of the true solution in some rigorous sense.

Notes

1. C denotes the set of complex numbers.
2. The discussion in this section is borrowed from [10].

3 NOISE MODELS

To reach the final goal of simulating and characterizing the effect of noise on the performance of an electronic circuit or system, we first need to investigate the actual noise sources in the system and develop models for these noise sources in the framework of the theory of signals and systems we will be operating with. The models we are going to use for the actual noise sources will be developed and described within the framework of the theory of probability and stochastic processes outlined in Chapter 2. In this chapter, we will first discuss the physical origins of noise sources in electronic systems. As mentioned in Chapter 1, we will concentrate on those noise sources which have their origin in the "random" statistical behavior of the atomic constituents of matter, and hence which are fundamentally unavoidable, i.e. the so-called *electrical* or *electronic* noise. We will exclude from our discussion those noise sources which have their origin in external effects on the system, e.g. atmospheric and power line interferences.

3.1 Physical Origins of Electrical Noise

Electrical noise is, basically a consequence of discrete or particle nature of matter and energy. Electrical charge is not continuous, but is carried in discrete amounts equal to the electron charge. Most *macroscopically* observable physical variables, such as electric current, are only *averages* over a large number of particles, e.g. electrons, of some parameter describing those particles [17]. When observed more precisely, the statistical or "random" nature of the macroscopic variables become apparent from the fluctuations in their values around the average. A complete specification of the microscopic state of a macroscopic system (such as a resistor) in classical physics would involve a specification of not only of the physical structure but also of the coordinates and momenta of all of the atomic and subatomic particles involved in its construction [18]. This is not possible, considering the *large* number of particles we are talking about. A complete specification on a particle by particle basis does not make sense in quantum mechanics, but the specification of an exact quantum state for the system as a whole is equally impossible [18]. Hence, the state of a macroscopic system is coarsely specified in terms of a relatively small number of parameters. A branch of physics, statistical mechanics, allows us to make predictions about the behavior of the system. The objective of statistical mechanics is to describe the properties of a large system in terms of macroscopic *state variables* such as temperature, volume, resistance, etc. One can calculate the most probable state of a system when it is subject to various constraints [19]. In particular, one can calculate the probability with which any possible microscopic configuration of particles will occur. Since each microscopic configuration corresponds to a particular value of the macroscopically observable quantities associated with the system, this allows us to calculate the probability distribution of one of these quantities, for example the current through a resistor. Although one can not specify the exact configuration of the particles at any instant, these particles are subject to precise dynamical laws, and these laws govern the evolution of one microscopic configuration into another as time proceeds [18]. One of the most significant and simplest system constraints is *thermodynamic equilibrium*. The concept of thermodynamic equilibrium is fundamentally associated with time-invariance. A system in thermodynamic equilibrium will remain in that state forever unless acted on by an external force. With different assumptions on the characteristics of particles in a system, and with constant mass and energy constraints, one arrives at different equilibrium distributions. For instance, Maxwell-Boltzmann statistics is for a system with classical particles, and Bose-Einstein and Fermi-Dirac statistics are for systems with quantum particles.

An ideal gas is a system of identical noninteracting particles. Although it is a classical system, its properties are useful in illustrating many basic physical principles. Many properties of conduction electrons in a metal, for example, can be derived from the assumption that they behave like an ideal gas. The ideal-gas law relates the macroscopic variables of the system.

$$pV = nkT \tag{3.1}$$

where p is the pressure, V is the volume, n is the number of particles, and T is the absolute temperature. The quantity k is a universal constant, called *Boltzmann's constant* (1.38×10^{-23} J/K). Starting with the ideal-gas law, and assuming that the system is in thermodynamic equilibrium (which means that the temperature of the gas and its container of volume V are the same, and hence collisions of the particles with the walls of the container are elastic, i.e no energy is exchanged in collisions), one can calculate the mean kinetic energy E of an ideal-gas particle to be

$$E = \frac{3}{2}kT. \tag{3.2}$$

The corresponding average *thermal velocity* corresponding to the above average kinetic energy for an electron has a value on the order of 10^7 cm/s. With the assumption that the ideal-gas particles do not interact and the gravity is negligible, the x, y and z components of motion are independent. Each particle is said to have three degrees of freedom. For any one degree i, we have

$$E_i = \frac{1}{2}kT. \tag{3.3}$$

The quantity $1/2kT$ is often referred to as the *thermal energy per degree of freedom*. "Random" thermal motion is a very general phenomenon that is a characteristic of all physical systems, and is the cause of so-called *thermal noise*. In the context of electronic systems, thermal noise is a result of the "random" thermal motion of *charge carriers* in a device [19].

Let us consider a fictitious electronic device with free charge carriers. The terminal currents of the device are caused by charge carriers which are drawn from a *source* or *emitter* of carriers in some region of the device. The voltages at the terminals of the device produce fields with which the above carriers interact in some *interaction region*, which gives rise to the particular terminal characteristics for that device. The fluctuations in the terminal voltages and currents arise from certain stochastic properties of the charge carriers. In the most general case, the stochastic properties of the carriers will be determined by the "source" region as well as the "interaction" region of the device. For thermal noise, the stochastic properties of the charge carriers are established in a region where the charge carriers are in thermal equilibrium. If there is no applied external electric field, the charge carriers are in thermal equilibrium with the crystal lattice (in much the same sense that ideal-gas molecules are in thermal equilibrium with the walls of a container) with an average velocity equal to the thermal velocity for the particular charge carrier and the temperature. When an external field is applied, the acceleration of the electrons does not result in a significant change in the velocity of the charge carriers between different collisons. The average component of the velocity added by the external field is called the *drift* velocity.

Now, we give a definition for the so-called *shot noise* using the notions that were introduced in the discussion of this fictitious electronic device. The noise present at the terminals of the device can be called *shot noise* when the stochastic properties of the charge carriers are determined by the "source" region of the device, and not by the interaction region. So, for shot noise, the region in which the charge carriers interact with external fields is physically distinct from the region in which their statistical properties are established. Thus, as they interact with external fields, they neither influence nor are influenced by random processes in the source region. For thermal noise, on the other hand, the interaction region coincides with the region where the carrier fluctuations are generated and, during their interaction, the carriers remain in approximate thermal equilibrium with a lattice. Still, thermal noise and shot noise are not mutually exclusive phenomena. Even for a device that has distinct source and interaction regions, if the charge carriers are in thermal equilibrium throughout, the noise can be considered to be of thermal origin. Then we can say that any classical system in thermal equilibrium is directly subject to the laws of thermodynamics, and the only noise it can exhibit is thermal noise. If a device is not in thermodynamic equilibrium, it may (and usually does) have other sources of noise[1].

3.1.1 Nyquist's theorem on thermal noise

In 1906, Einstein predicted that Brownian motion of the charge carriers would lead to a fluctuating e.m.f. across the ends of any resistance in thermal equilibrium. The effect was first observed by Johnson (1928), and its spectral density was calculated by Harry Nyquist (1928).

The first fundamental result in the theory of thermal noise is the Nyquist theorem. We will now present a restricted version of the statement for the theorem and describe its generalizations later. Nyquist's theorem states that the random fluctuations in the short-circuit terminal current of an arbitrary linear resistor (as a matter of fact, any two-terminal linear electrical circuit with a purely resistive impedance), having a resistance R, maintained in thermal equilibrium at a temperature T, are independent of such parameters as its conduction mechanism, composition, construction, or dimensions, and depend only upon the values of the resistance R and the temperature T, and can be modeled as a WSS stochastic process with a spectral density given by

$$S_{th,i}(f) = \frac{2kT}{R} \qquad (3.4)$$

where k is Boltzmann's constant. This is the spectral density for a white noise process. To be consistent with our previous definition of spectral density (which is defined for negative frequencies as well as positive frequencies), a factor of 2 appears in the above formula as opposed to the usual form with a factor of 4. The statement for the theorem can be also stated for the open-circuit terminal noise voltage. We are not going to present a derivation for Nyquist's result. See [17] for a discussion of different methods of derivation for Nyquist's theorem. There are various generalizations of the theorem, from an arbitrary interconnection of impedances to nonreciprocal linear networks. Nyquist's result applies to any system in which the charge carriers are in equilibrium with a lattice, and can be assigned a temperature T equal to that of the lattice. See [17] for a discussion of the generalizations of Nyquist's theorem.

The spectral density given in (3.4) is the one for a white noise. This, of course, can not be true for a physical noise process. A quantum-statistical calculation (which was considered by Nyquist in his original treatment) of the spectral density yields

$$S_{th,i}(f) = 2R(\frac{1}{2}hf + \frac{hf}{\exp(\frac{hf}{kT}) - 1}) \qquad (3.5)$$

where h is Planck's constant. In the limit $hf/kT \to 0$, (3.5) reduces to (3.4). (3.5) can be expressed as

$$S_{th,i}(f) = 2kTR(\frac{1}{2}f/f_0 + \frac{f/f_0}{\exp(f/f_0) - 1}) \qquad (3.6)$$

where $f_0 = 6000$GHz at room temperature. Hence, for practical purposes, at room temperature, (3.5) is very well approximated by (3.4). Practical electronic systems have bandwidths much smaller than 6000GHz.

It can be shown that the thermal noise of a linear resistor as a white WSS stochastic process with spectral density (3.4) is accurately modeled by a *Gaussian* process (derived by G.E Uhlenbeck and L.S. Ornstein in 1930), as a direct consequence of the central limit theorem. Thus, as a white Gaussian process, we identify the model for thermal noise of a linear resistor as the formal derivative of the Wiener process.

The thermal noise model based on Nyquist's theorem for a linear resistor is very convenient for use in circuit theory. A noisy linear resistor is modeled as a noiseless linear resistor with the same resistance R, and a noise current source connected across the terminals of the noiseless resistor representing a WSS white Gaussian stochastic process with spectral density given by (3.4). This is a Norton equivalent model for the noisy resistor. Similarly, in a Thevenin equivalent model, a noise voltage source is connected in series with the noiseless resistor. The noise voltage source also represents a WSS white Gaussian process with a spectral density

$$S_{th,v}(f) = 2kTR. \tag{3.7}$$

With the above models of thermal noise of a linear resistor in circuit theory, it is implicitly assumed that the internal noise generation for the resistor is independent of the load connected across the resistor. At this point, we would like to reemphasize the fact that Nyquist's result was derived for a *linear* resistor in *thermal equilibrium*.

3.1.2 Shot noise

We have already tried to give an "intuitive" explanation for shot noise in general and its distinction from thermal noise. Now, we discuss shot noise in semiconductor devices. In vacuum tubes (where shot noise was first observed) and semiconductor devices incorporating a *pn* junction, the charge carriers interact with fields due to applied external voltages only in specific regions of the device, e.g. the depletion layer at the junction. The reason for this is simply that the fields within the devices, due to applied voltages, are approximately zero except in these regions. A *pn* junction consists of two contacting regions of relatively highly doped material, of two different types, both regions being fitted with ohmic contacts. Except for a small thermal noise contribution due to their finite conductance, the bulk p and n regions make no contribution to the noise. In a depletion layer for a *pn* junction, the motion of the electrons is dominated by the effect of the macroscopic field in the layer, and is little influenced by the very "small" number of collisions that occur in the layer. Therefore, the fluctuations in the behavior of the electrons in the interaction region are due to the fluctuations in the *emissions* into the region, i.e. *shot noise*. For instance, in a bipolar transistor, the fluctuations in the active carrier flow are established within the emitter and, to a lesser extent, within the base where collisions occur but there is approximately no field due to externally applied voltages. Fields and interactions occur only in the emitter-base and base-collector depletion layers. Hence, bipolar transistors are typical *shot noise limited* devices. By contrast, in a field-effect transistor, the carrier fluctuations are established in the channel, and this is also where the carriers interact with the applied voltages. A field-effect transistor is, therefore, a *thermal noise limited* device [18]. For MOS field-effect transistors in weak-inversion (subthreshold operation), the generated noise can also be explained using shot noise arguments. In fact, at thermal equilibrium, both shot and thermal noise arguments produce identical results, consistent with the previous comment on shot and thermal noise not being mutually exclusive [20]. A similar observation can also be made for a *pn* junction at thermal equilibrium, which we will present later in this chapter.

A stochastic process model for shot noise will be presented later in this chapter. At this point, we will only state that the shot noise in the current through a *pn* junction which is biased by a *time-invariant* signal can also be modeled as a WSS white Gaussian process for practical purposes.

3.1.3 Flicker or $1/f$ noise

The noise at "low" frequencies for semiconductor devices often considerably exceeds the value expected from thermal and shot noise considerations. The spectral density of noise increases as the frequency decreases, which is observed experimentally. Generally, this low frequency excess noise is known as *flicker noise*, and in many cases the spectral density is inversely proportional to the frequency. The characteristics of flicker noise considerably change from device to device, even for two same type of devices on the same die. This suggests that flicker noise is often associated with the fine details of the device structure [18]. Flicker noise has been experimentally observed in a great variety of electronic components and devices, including carbon resistors, *pn* junction diodes, bipolar transistors, and field-effect transistors. Unfortunately, the precise mechanisms involved in flicker noise are complicated, vary greatly from device to device, and have been the subject of speculation and controversy [20]. For bipolar transistors, it is believed to be caused mainly by traps associated with contamination and crystal defects in the emitter-base depletion layer. These traps capture and release carriers in a random fashion and the time constants associated with the process give rise to a noise signal with energy concentrated at low frequencies [1]. For MOS field-effect transistors, according to some studies, the carrier density fluctuations caused by the exchange of charge carriers between the channel and the interface traps are the cause [20]. Many theories have been proposed for flicker noise in various electronic components. Each theory is involved, and gives rise to a different spectral density expression for the stochastic process that models the noise current in the device. The one important point on which all the theories, as well as experimental results, agree is that the spectral density for the stochastic process that models flicker noise is approximately inversely proportional to the frequency f for a WSS stochastic process model [20]. This is the reason why flicker noise is also referred to as $1/f$ *noise*. At this point, we would like to emphasize the fact that any device in thermal equilibrium is directly subject to the laws of thermodynamics, so its noise is solely thermal noise modeled as a WSS white Gaussian process.

Thus, flicker or $1/f$ noise can only occur in *non-equilibrium* situations in devices subjected to applied bias voltages or bias voltages derived from a signal [18].

$1/f$ noise has been observed as fluctuations in the parameters of many systems apart from semiconductor devices. Many of these systems are completely unrelated to semiconductors. $1/f$ type fluctuations have been observed in average seasonal temperature, annual amount of rainfall, rate of traffic flow, economic data, loudness and pitch of music, etc. [21]. Keshner in [21] describes $1/f$ noise as

> $1/f$ noise is a nonstationary stochastic process suitable for modeling evolutionary or developmental systems. It combines the strong influence of the past events on the future and, hence somewhat predictable behavior, with the influence of random events.

The presence of $1/f$ noise in such a diverse group of systems has led researchers to speculate that there exists some *profound* law of nature that applies to all nonequilibrium systems and results in $1/f$ noise. Keshner in [21] "speculates"

> The fact that music has $1/f$ noise statistics and that when notes are chosen at random, they sound most musical when their spectral density is $1/f$, suggests a connection between the way humans perceive and remember, and the structure of $1/f$ noise. Because of this connection and the influence of human memory and behavior on the development of our institutions: the development of ourselves, our economic system, our government, and our culture may each have the statistics of a $1/f$ noise stochastic process.

3.2 Model for Shot Noise as a Stochastic Process

For a *pn* junction diode, it can be shown that the current in the external leads is, approximately, a result of the charge carriers crossing the depletion layer [18]. For every charge carrier crossing the depletion layer, the current in the external leads consists of a short pulse, of total charge q of the charge carrier. Now, let us assume that the charge carriers are being emitted into the depletion layer from only one side of the junction. Then, let $N(t)$ be the number of charge carriers that have crossed the depletion layer prior to time t, with the initial condition $N(0) = 0$. Assuming that the carrier crossings are instantaneous, $N(t)$ can be "reasonably" modeled with an inhomogeneous Poisson counting process with time-varying rate $\lambda(t)$ (see Section 2.2.8). For a constant rate (homogeneous) Poisson counting process, we recall that inter-crossing times for the carriers are exponentially distributed, the only continuous distribution that has the memoryless property described in Example 2.2. Then, the stochastic process that describes the total charge that has crossed the depletion layer prior to time t is given by

$$Q(t) = qN(t) \qquad (3.8)$$

where q is the charge of one charge carrier. The stochastic process that describes the current through the junction is, by definition, given as the time derivative of $Q(t)$. However $Q(t)$ has instantaneous jumps exactly at the times of the instantaneous carrier crossings, and it is *not* mean-square differentiable. From Section 2.2.10, we know that the formal derivative of a Poisson counting process (after mean is subtracted) can be identified as a white noise process. Then, in the generalized sense of differentiation for stochastic processes, the formal time derivative of $Q(t)$ can be expressed as

$$I(t) = \frac{d}{dt}Q(t) = \sum_i q\delta(t - T_i) \qquad (3.9)$$

where T_i are the (random) crossing times for the Poisson counting process $N(t)$. T_i are called a set of *Poisson time points*, and $I(t)$ in (3.9) is called a *Poisson pulse train* [7]. In (3.9), we let $-\infty < t < \infty$ and let the starting time for the Poisson counting process $N(t)$ go to $-\infty$ so that the current process $I(t)$ is in "steady-state". In reality, the charge carrier crossings can not be instantaneous. Now, we replace the δ function in (3.9) with a finite width pulse $h(t)$ which also has a total area equal to 1, and satisfies $h(t) = 0$ for $t < 0$. This is equivalent to passing the Poisson pulse train in (3.9) through an LTI system with impulse response $h(t)$ [7]. We get

$$I(t) = \sum_i qh(t - T_i). \qquad (3.10)$$

The shape of the pulse $h(t)$ is determined by the specific characteristics of the device. As far as we are concerned, it is a pulse with area equal to 1 and the pulse width is equal to the *transit time* of the carrier through the depletion layer. One can calculate the mean of the current $I(t)$ [7], using the properties of the Poisson counting process, to be

$$\bar{I}(t) = \mathbf{E}\left[I(t)\right] = q \int_{-\infty}^{\infty} \lambda(u)h(t-u)du. \qquad (3.11)$$

Now, we define the noise current $\tilde{I}(t)$ to be the difference between the total current $I(t)$ and its mean $\bar{I}(t)$

$$\tilde{I}(t) = I(t) - \mathbf{E}\left[I(t)\right]. \qquad (3.12)$$

Obviously, the mean of the noise current $\tilde{I}(t)$ is zero, and its autocorrelation (which is equal to the autocovariance of the total current $I(t)$) can be calculated to be [7]

$$R_{\tilde{I}}(t,\tau) = \mathbf{E}\left[\tilde{I}(t+\tau/2)\tilde{I}(t-\tau/2)\right] \qquad (3.13)$$

$$= q^2 \int_{-\infty}^{\infty} \lambda(u) h(t+\tau/2-u) h(t-\tau/2-u) du. \qquad (3.14)$$

If the rate $\lambda(t) = \lambda$ is a constant, then (3.11) and (3.14) reduce to

$$\bar{I}(t) = \bar{I} = q\lambda \int_{-\infty}^{\infty} h(t-u) du = q\lambda \qquad (3.15)$$

and

$$\begin{aligned} R_{\tilde{I}}(t,\tau) &= q^2 \lambda \int_{-\infty}^{\infty} h(t+\tau/2-u) h(t-\tau/2-u) du & (3.16) \\ &= q^2 \lambda \int_{-\infty}^{\infty} h(u+\tau/2) h(u-\tau/2) du & (3.17) \\ &= R_{\tilde{I}}(\tau). & (3.18) \end{aligned}$$

From (3.15) and (3.17), we conclude that the shot noise current $\tilde{I}(t)$ is a WSS process for constant rate λ. The spectral density of $\tilde{I}(t)$ with a constant rate λ can be calculated to be [7]

$$S_{\tilde{I}}(f) = \lambda q^2 |H(f)|^2 \qquad (3.19)$$

where $H(f) = \mathbf{F}\{h(t)\}$. Note that $H(0) = 1$, since the area under $h(t)$ is equal to 1. From (3.15), we have

$$\lambda = \frac{\mathbf{E}[I(t)]}{q} = \frac{\bar{I}(t)}{q} = \frac{\bar{I}}{q}. \qquad (3.20)$$

If we substitute (3.20) in (3.19), we get

$$S_{\tilde{I}}(f) = q\bar{I}|H(f)|^2, \qquad (3.21)$$

which expresses the shot noise current spectral density in terms of the mean current and the charge of a carrier. If $h(t)$ is the Dirac delta function, then (3.21) reduces to

$$S_{\tilde{I}}(f) = q\bar{I} \qquad (3.22)$$

which is the spectral density for a WSS white noise process! This is no surprise to us, because we have already seen in Section 2.2.10 that the formal derivative of a Poisson counting process with a constant rate (after the mean is subtracted) can be identified as a WSS white noise process. The "usual" factor "2" does not appear in (3.22), because the spectral density in (3.22) is a double-sided density, i.e. it is defined for negative frequencies as well.

We go back to the shot noise process with the time-varying rate $\lambda(t)$. Now, we assume that the change in $\lambda(t)$ as a function of time is *slow enough* so that it can be approximated with a constant over the width of a single current pulse (i.e. the width of $h(t)$). Then, we can approximate (3.11) as

$$\bar{I}(t) = \mathbf{E}\left[I(t)\right] = q\int_{-\infty}^{\infty} \lambda(u)h(t-u)du \qquad (3.23)$$

$$\approx q\int_{-\infty}^{\infty} \lambda(t)h(t-u)du \qquad (3.24)$$

$$\approx q\lambda(t)\int_{-\infty}^{\infty} h(t-u)du \qquad (3.25)$$

$$\approx q\lambda(t) \qquad (3.26)$$

where we have used the fact that the area under the pulse $h(t)$ is 1. Now, let the pulse width of $h(t)$ be τ_{tr}, which is equal to the transit time of carriers crossing the depletion layer. From (3.14), $R_{\tilde{I}}(t,\tau) = 0$ for $|\tau| > \tau_{tr}$. For $|\tau| \leq \tau_{tr}$ (as a matter of fact for all τ), we can approximate (3.14) as

$$R_{\tilde{I}}(t,\tau) = \mathbf{E}\left[\tilde{I}(t+\tau/2)\tilde{I}(t-\tau/2)\right] \qquad (3.27)$$

$$= q^2 \int_{-\infty}^{\infty} \lambda(u)h(t+\tau/2-u)h(t-\tau/2-u)du \qquad (3.28)$$

$$\approx q^2 \int_{-\infty}^{\infty} \lambda(t)h(t+\tau/2-u)h(t-\tau/2-u)du \qquad (3.29)$$

$$\approx q^2\lambda(t)\int_{-\infty}^{\infty} h(t+\tau/2-u)h(t-\tau/2-u)du \qquad (3.30)$$

$$\approx q^2\lambda(t)\int_{-\infty}^{\infty} h(u+\tau/2)h(u-\tau/2)du. \qquad (3.31)$$

Then using (3.31), the instantaneous (time-varying) spectral density (see Section 2.2.6) of $\tilde{I}(t)$ can be calculated to be

$$S_{\tilde{I}}(t,f) = \lambda(t)q^2|H(f)|^2 \qquad (3.32)$$

where $H(f) = \mathbf{F}\{h(t)\}$. From (3.26), we have

$$\lambda(t) = \frac{\mathbf{E}\left[I(t)\right]}{q} = \frac{\bar{I}(t)}{q}. \qquad (3.33)$$

If we substitute (3.33) in (3.32), we get

$$S_{\tilde{I}}(t,f) = q\bar{I}(t)|H(f)|^2 \qquad (3.34)$$

which expresses the shot noise current (time-varying) spectral density in terms of the (time-varying) mean current and the charge of a carrier. (3.26) and (3.34) seem to be straightforward generalizations of (3.15) and (3.21) obtained by simply replacing the constant rate λ in (3.15) and (3.21) by the time-varying rate $\lambda(t)$, but these generalizations are correct only for the case where the change in $\lambda(t)$ as a function of time is slow enough so that it can be approximated by a constant value over the width of $h(t)$.

If $h(t)$ is the Dirac impulse, then (3.34) reduces to

$$S_{\tilde{I}}(t,f) = q\bar{I}(t) \tag{3.35}$$

which is independent of f, and hence, is similar to the spectral density of a WSS white noise process which has a spectral density independent of f. However, there is one important difference: The stochastic process that has the spectral density in (3.35) is *not* a WSS process, because its spectral density is *not* a constant function of t.

Up to this point, we have discussed the second-order probabilistic characterization, i.e the autocorrelation and spectral density, of shot noise. Now, we discuss the fdds for the shot noise process. It can be shown that if the rate $\lambda(t)$ is large compared with the inverse of the width of the pulse $h(t)$, then the shot noise process $\tilde{I}(t)$ is approximately *Gaussian* [7]. If $\lambda(t)$ is large compared with the inverse of the width of the pulse $h(t)$, it means that many pulses overlap at each instant in time, and hence the central limit theorem can be used to conclude that the shot noise process is approximately Gaussian [5]. For instance, if the mean current $\bar{I}$ is 1mA, and if the charge carriers are electrons with charge $q = 1.6 \times 10^{-19}$C, then the rate λ

$$\lambda = \frac{\bar{I}}{q} = 6.25\ 10^{15}\ 1/\text{sec} \tag{3.36}$$

is much greater than the inverse of $\tau_{tr} = 10^{-12}$sec, which is a typical transit time (i.e. pulse width) for a *pn* junction diode.

As a Gaussian process, and as a WSS white noise process with spectral density given by (3.22), the shot noise process $\tilde{I}(t)$ with a constant rate λ can be identified as the formal derivative of the Wiener process, i.e. $\xi(t)$ (up to a scaling factor to be discussed) defined in Section 2.2.7. As a matter of fact, it can be shown that the spectral density of the Gaussian WSS process $\sqrt{q\bar{I}}\xi(t)$ is exactly given by (3.22). Hence, as Gaussian WSS processes having the same mean and the spectral density, the shot noise process $\tilde{I}(t)$ with a constant rate λ, and the scaled version of the derivative of the Wiener process $\sqrt{q\bar{I}}\xi(t)$ are "probabilistically equivalent", i.e. they have the same fdds. Similarly, it can be shown that the time-varying spectral density of the Gaussian, but not necessarily WSS, process $\sqrt{q\bar{I}(t)}\xi(t)$ is exactly given by (3.35). Hence, as Gaussian processes having the same mean and the time-varying spectral density, the shot noise process $\tilde{I}(t)$ with a time-varying rate $\lambda(t)$, and the "modulated" version of the derivative of the Wiener process $\sqrt{q\bar{I}(t)}\xi(t)$ are "probabilistically equivalent", i.e. they have the same fdds. In the above discussion, we have assumed that $q\bar{I}(t) \geq 0$. Obviously, if the charge carriers are electrons, we have $q < 0$ and $\bar{I}(t) < 0$.

We conclude that the shot noise with a time-varying rate can be modeled as a modulated version of the derivative of the Wiener process

$$\tilde{I}(t) = \sqrt{q\bar{I}(t)}\xi(t) \tag{3.37}$$

This result was achieved with the following three basic assumptions:

- The emission of carriers into the depletion layer can be "reasonably" modeled as a Poisson process.

- The change in the time-varying rate $\lambda(t) = \bar{I}(t)/q$ as a function of time is slow enough so that it can be approximated by a constant value over the width of a current pulse $h(t)$ caused by a single charge carrier.

- The time constants of the system (the shot noise process is affecting) are much larger than the carrier transit time, so that the finite width pulse $h(t)$ can be approximated by an ideal impulse function.

Of course, these last two assumptions are closely related, and express the same condition from two different points of view. The connection between these assumptions becomes clear when one considers the fact that the mean current $\bar{I}(t)$ (which sets the time-varying rate as $\lambda(t) = \bar{I}(t)/q$ for the shot noise process) is set by the system that the *pn* junction is a part of.

Shot noise is not a phenomenon specific to *pn* junctions. Whenever a device contains a potential barrier (e.g. *pn* junctions, vacuum tubes, MOSFETs in weak inversion, etc.), the current is limited to those charge carriers that have sufficient energy to surmount the potential barrier. In our above discussion of the *pn* junction to derive a model for shot noise as a stochastic process, we have assumed that there is only one kind of charge carrier present, and the charge carriers are emitted from only one side of the depletion layer. The real situation is more complicated, of course. There are two types of charge carriers, electrons and holes, and they are emitted into the depletion layer from both sides of the junction. Now, we summarize Robinson's discussion from [18] for the different current components for a *pn* junction diode. To simplify the discussion, we assume that recombination between holes and electrons is negligible. Hence, electron and hole currents can be considered to be independent and additive. (The summation of two independent Poisson counting processes is another Poisson counting process, and the rate for the sum counting process is given by the summation of the rates for the two Poisson counting processes that are being added.) Now, consider the electrons. The electron current in a *pn* junction consists of two components. A forward component consisting of electrons from the n region crossing to the p region, and a reverse component passing from p to n. At zero bias across the junction, these currents balance on average. The reverse current is controlled by density of electrons (minority carriers) in the p material, and there is no potential barrier to their transit into the n region. This current is unchanged by an applied bias, which is denoted by V. The forward current is controlled partially by the majority carrier (electron) concentration in the n material, but, since this is large, the dominant effect is due to the retarding potential barrier at the junction. An applied potential V, with the p region positive, reduces this barrier, and the forward current increases by a factor $\exp(V/V_T)$, where $V_T = kT/q$ with Boltzmann's constant k, the temperature T and the electron charge q. So, the total current I is given by

$$I = I_s(\exp(V/V_T) - 1). \tag{3.38}$$

Each component of I (modeled as the derivative of independent Poisson processes) displays shot noise, so the shot noise current for the diode can be modeled with a Gaussian process with possibly time-varying spectral density given by

$$S(t,f) = qI_s(\exp(V/V_T) + 1) = q(I + 2I_s). \tag{3.39}$$

The differential or small-signal conductance of the diode is

$$G(V) = \frac{dI}{dV} = \frac{I_s}{V_T}\exp(V/V_T). \tag{3.40}$$

Then, (3.39) can be expressed as

$$S(t,f) = 2kTG(V) - qI(V). \tag{3.41}$$

At zero bias, $V = 0$, (3.41) becomes

$$S(t,f) = 2kTG(0). \tag{3.42}$$

At zero bias, the diode is a system in thermodynamic equilibrium, so (3.42) can be interpreted as the spectral density of thermal noise associated with the conductance $G(0)$ [18]. However, note that, the conductance $G(0)$ *does not* correspond to a physical resistor. A similar observation for the noise of a MOSFET in weak inversion can be made [20].

The shot noise model derived above for a time-varying bias current $\bar{I}(t)$ as a modulated version of the derivative of the Wiener process is very convenient for use in circuit theory to model shot noise associated with various current components of devices such as *pn* junction diodes, bipolar junction transistors, etc. Shot noise sources in these devices are modeled with noise current sources connected across various nodes of the devices, and they represent Gaussian stochastic processes with spectral density given by (3.35). For instance, shot noise in a bipolar transistor is modeled with two *independent* current noise sources, which are connected across the base-emitter and collector-emitter junctions.

It is interesting to observe that all the devices that display shot noise (*pn* junction diodes, bipolar transistors, vacuum tubes, MOSFETs in weak inversion) obey the same exponential relationship between the terminal currents and voltages. The explanation of this goes back to the first principles of thermodynamics, i.e. Maxwell-Boltzmann statistics, and the Second Law of Thermodynamics [22]. Coram and Wyatt in [22] prove a *nonlinear fluctuation-dissipation theorem* for devices that can be accurately described by a model given by (3.38), where the forward and reverse currents satisfy

$$\frac{I_f}{I_r} = \exp(V/V_T). \tag{3.43}$$

Finally, we would like to emphasize the fact that shot noise is not only a result of the fact that electrical charge is carried in discrete amounts equal to the electron charge, but also a result of the fact that charge carriers which attain sufficient energy to overcome the potential barrier enter the depletion layer *randomly*, as modeled by a Poisson process. If the emission of charge carriers was spaced *evenly in time* instead of *randomly*, the current would be a deterministic periodic waveform which obviously does not have a white spectral density.

3.3 Model for Thermal Noise as a Stochastic Process

In Section 3.1.1, we have seen that the thermal noise of a linear resistor in thermodynamic equilibrium can be modeled as a WSS Gaussian white current noise source with spectral density given by

$$S_{th,i}(f) = \frac{2kT}{R}. \tag{3.44}$$

As a Gaussian process, and as a WSS white noise process, the thermal noise of a linear resistor R in thermodynamic equilibrium can be identified as the formal derivative of the Wiener process, i.e. $\xi(t)$ (up to a scaling factor to be discussed) defined in Section 2.2.7. As a matter of fact, it can be shown that the spectral density of the Gaussian WSS process $\sqrt{2kT/R}\,\xi(t)$ is exactly given by (3.44). Hence, as Gaussian WSS processes having the same mean (which is zero) and spectral density, the thermal noise current process and the scaled version of the derivative of the Wiener process $\sqrt{2kT/R}\,\xi(t)$ are "probabilistically equivalent", i.e. they have the same fdds.

In Section 3.2, we derived a stochastic process model for shot noise in devices biased with possibly time-varying signals, which was based on several assumptions stated in Section 3.2. Now, the question is: Can Nyquist's theorem be generalized to nonlinear dissipative systems biased with possibly time-varying signals? A nonlinear resistor with a time-varying bias is, obviously, not a system in thermodynamic equilibrium. It turns out that the above question has been a widely explored research topic in both physics and engineering literature. Gupta in [17] summarizes the work done in the area of the theory of thermal noise in nonlinear dissipative systems up to 1982. The general theory of thermal noise in nonlinear dissipative systems is quite intricate, but from an *engineering* perspective the problem can be stated as: Can we use a straightforward generalization of Nyquist's theorem for nonlinear resistors with a time-varying bias, simply by replacing the R in (3.44) with the time-varying small-signal resistance $R(t)$ obtained by differentiating the nonlinear relationship that relates the terminal voltages and currents for the particular nonlinear dissipative device and evaluating it at a trajectory set by the time-varying bias? This generalization would be similar to the one we described for shot noise. The answer to the above question is *no* in the general case as explained by Gupta in [17]. Hence, the question is: Under what conditions can we make the above generalization, if any at all? "Intuitively", one might think that if the change in the time-varying bias as a function of time is "slow enough" so that the device stays approximately at thermal equilibrium, and the fluctuations caused by the noise are much "smaller" compared with the deterministic desired signals in the system, the above generalization is reasonable. Indeed this seems to be the case, i.e. Nyquist's theorem applies to systems in approximately thermal equilibrium with the understanding that the resistance is the small-signal resistance of the system [17]. However it is stated in [17] that this result is not rigorously established (probably as a result of the imprecise statement of the assumptions and approximate nature of the theory, we believe), and is not universally valid. Nyquist's theorem has been *liberally* applied to nonlinear dissipative devices in electronic device literature to develop noise models, the most conspicuous one being the MOSFET for a model of its channel noise. The results obtained with the noise models arrived at by applying Nyquist's theorem to biased nonlinear dissipative devices seem to match experimental results. However, in our opinion, one has to be very suspicious of measurements matching results obtained with the noise models. Because, in noise measurements, some of the parameters describing the noise model are usually "calibrated" using some measurement results. So, it is not always clear that the match is due to a good "calibration" of the parameters, a reasonable noise model, or both. This might be acceptable when one has access to the devices, as stand-alone components, she/he is going to use for her/his design. This is usually the case for microwave design. However, in IC design, the particular device that is important for the noise performance may not be, and almost always is not, accessible. The parameters describing the noise behavior of a device can be quite different for two devices from the same fab, and even for the devices on the same die, or on the same chip.

In summary, a rigorous justification for any thermal noise model used for a device, which does not satisfy the assumptions of the Nyquist theorem, is definitely needed. Recent work by Coram and Wyatt [22] focuses on testing the consistency of these "unrigorously" generalized noise models with basic equilibrium (Maxwell-Boltzmann statistics) and non-equilibrium (monotonically increasing entropy as the system relaxes to equilibrium) thermodynamic principles. Using techniques from the theory of stochastic differential equations, they conclude that, in the general case, these "generalized" thermal noise models for nonlinear dissipative devices do *not* yield results that are consistent with basic principles of thermodynamics.

From a practical "engineering" perspective, modeling the thermal noise of a nonlinear dissipative device as a Gaussian stochastic process obtained by "modulating" the derivative of a Wiener process seems to be reasonable with the following assumptions:

- The change in the time-varying bias as a function of time is "slow enough" so that the nonlinear dissipative device stays approximately at thermal equilibrium.

- The fluctuations caused by the noise are much "smaller" compared with the deterministic desired signals in the system.

The above statement, of course, is not rigorously justified.

Let $G(t)$ be the time-varying small signal conductance obtained by differentiating the nonlinear relationship between the terminal currents and voltages of a dissipative device evaluated at a time-varying trajectory imposed by the deterministic large signals in the system. Note that the nonlinear relationship has to describe a *dissipative* device. For instance, the small-signal input resistance r_π of a bipolar transistor does not correspond to a physical resistor, hence does not generate thermal noise [1]. The thermal noise of the device is modeled as a current noise source across the terminals of the device, and it is mathematically described as a modulated version of the derivative of the Wiener process, i.e $\sqrt{2kTG(t)}\xi(t)$, and the time-varying spectral density for this *nonstationary* Gaussian process is given by

$$S_{th,i}(t,f) = 2kTG(t) \qquad (3.45)$$

3.4 Models for Correlated or non-White Noise

In the two previous sections, we concluded that, for practical purposes, thermal noise and shot noise in electronic devices and components can be modeled as modulated versions of a white Gaussian process, i.e. by modulating the derivative of the Wiener process. The Gaussian processes that represent *current* noise sources connected across the two terminals of a device and modeling thermal and shot noise have time-varying spectral densities given by

$$S_{thermal}(t,f) = 2kTG(t) \qquad (3.46)$$
$$S_{shot}(t,f) = q\bar{I}(t). \qquad (3.47)$$

These spectral densities correspond to the following autocorrelation functions

$$R_{thermal}(t,\tau) = 2kTG(t)\delta(\tau) \qquad (3.48)$$
$$R_{shot}(t,\tau) = q\bar{I}(t)\delta(\tau). \qquad (3.49)$$

The modulated processes which have the spectral densities and the autocorrelation functions given above are expressed as

$$\tilde{I}_{thermal}(t) = \sqrt{2kTG(t)}\xi(t) \qquad (3.50)$$
$$\tilde{I}_{shot}(t) = \sqrt{q\bar{I}(t)}\xi(t) \qquad (3.51)$$

where $\xi(t)$ is the WSS white Gaussian process, i.e. the derivative of the standard Wiener process. The stochastic processes defined by (3.50) and (3.51) are *not* WSS processes, but as seen from (3.48) and (3.49), they are "delta-correlated", i.e. the τ variable appears only as the argument of a delta function in their autocorrelation functions, and hence the time-varying spectral densities (3.46) and (3.47) are *independent* of f. We will refer to such processes as *nonstationary white noise processes*.

The stochastic process models summarized above for thermal noise and shot noise are exactly in the same form: modulated versions of the derivative of the Wiener process. Moreover, the modulation functions are fully described in terms of the device equations that relate the terminal voltages and currents, and these processes represent current noise sources to be connected across the nodes of the device. We will later see that the differential equations describing an electronic circuit containing thermal and shot noise sources modeled as above can be formulated as a system of *stochastic differential equations* (see Section 2.5), where the noise sources are *additively* superimposed on a system of ordinary differential equations that describe the deterministic or systematic behavior of the electronic circuit. This is rather desirable, because it enables us to use results from the theory of stochastic differential equations in analyzing the effect of noise sources on electronic circuits and systems. Thus, we would like to have models for other types of noise sources which satisfy the following two conditions:

- The only way "randomness" enters a model of any noise source has to be through the derivative of the Wiener process.

- In the formulation of differential equations describing the electronic circuit, the noise source model can be "additively" superimposed on top of the equations that describe the systematic behavior of the circuit.

These conditions might seem to be too restrictive at first sight, but models for most of the noise sources encountered in practice can be developed to satisfy these conditions. This will become clearer with the following observations: The model for the noise source can involve any number of Wiener processes (actually its derivative, the WSS Gaussian white noise process) which can be independent or correlated. Moreover, the noise model itself can contain a subsystem, i.e. the noise source can be modeled to be the output of a subsystem that has WSS Gaussian white noise inputs. Including a subsystem in the noise model usually requires the introduction of extra variables to describe the whole system, i.e. the state variables of the system have to be augmented. *State augmentation* is a common technique used in many problems. We will see examples for this later.

As a matter of fact, the modulating functions in the models of thermal and shot noise can be thought to be *memoryless time-varying*, but not necessarily linear, subsystems (see Section 2.3). If $G(t)$ in (3.46) and $\bar{I}(t)$ in (3.47) are functions of only time, then these subsystems are linear, but if $G(t)$ and $\bar{I}(t)$ also depend on some state variables (e.g. terminal voltages of the device) of the system, then these subsystems are *nonlinear*. Since they are memoryless, these subsystems do not have internal state variables, and hence do not require the introduction of extra state variables into the overall system. For a more "realistic" model of thermal noise or shot noise, one may consider introducing internal state variables to these subsystems to model a finite response time, for instance, between the changes in $\bar{I}(t)$ in (3.47) as a function of time and the appearance of the effect of these changes on the spectral density of shot noise.

For example, if we would like to model a noise source as a WSS Gaussian process with spectral density $S(f)$, this noise model can be realized as the output of a SISO, stable, LTI subsystem with the transfer function $H(f)$ satisfying

$$S(f) = |H(f)|^2 \qquad (3.52)$$

and setting the input to $\xi(t)$, the WSS Gaussian white noise. In general, this subsystem has internal state variables and requires state augmentation.

If we use a *linear* multi-input single-output (MISO) subsystem, with a number of WSS *Gaussian* white noise inputs, then the stochastic process at the output that models a noise source will also be *Gaussian*, but not necessarily WSS. If we would like to realize a *non-Gaussian* noise model, this can be accomplished with a *nonlinear* subsystem.

3.4.1 $1/f$ noise model for time-invariant bias

As discussed in Section 3.1.3, at present there does not exist a *unified theory* for $1/f$ noise. Using the results of experimental work (i.e. measurements), for practical purposes, $1/f$ noise in electronic devices and components associated with a direct time-invariant current I is modeled as a "WSS" stochastic process with spectral density given by

$$S_{1/f}(f) = K \frac{I^a}{f^b} \qquad (3.53)$$

where K is a constant for a particular device, a is a constant in the range 0.5 to 2, b is a constant approximately equal to 1, and f is the frequency. If $b = 1$, the spectral density has a $1/f$ frequency dependence. Observe that when $b = 1$, (3.53) can not be the spectral density of a "well-defined" stochastic process, because the variance of this WSS process has to be ∞, which can be observed using (2.47). Keshner in [21] argues that this problem arises, because $1/f$ noise is inherently a *nonstationary* process and should be modeled as such. Please see [21] for an in-depth discussion of this issue. Nevertheless, for "engineering" calculations, the model in (3.53) seems to be adequate for a time-invariant bias, but, as we will discuss later, the generalization of (3.53) to *time-varying* bias is not straightforward.

$1/f$ noise as modeled by (3.53) is not a "delta-correlated" process, and hence does not have independent values at every time point (unlike the thermal and shot noise models). Actually, its present behavior is *equally* correlated with both the recent and distant past [21]. Based on our previous discussion, we would like to develop a model for $1/f$ noise as the output of a subsystem with WSS white Gaussian processes as inputs. So, the question is: Does there exist a stable, SISO, LTI system, the output of which has the spectral density given by (3.53) (with $b = 1$ which is most common) when driven by a WSS white noise? This requires that the system transfer function $H(f)$ satisfies

$$|H(f)|^2 = S_{1/f}(f) = K\frac{I^a}{f}. \tag{3.54}$$

Keshner answers this question in [21]. The system he describes consists of a WSS white noise current source driving the input of a one-dimensional continuous resistor-capacitor (RC) line of infinite length. The transfer function (i.e. the impedance) $Z(f)$ of an infinite RC line is given by

$$Z(f) = (\frac{R}{j2\pi f C})^{1/2} \tag{3.55}$$

where R is the resistance of the line per unit length, and C is the capacitance of the line per unit length. Hence it satisfies (3.54) (up to a scaling), but this is true only if the line is *infinite*. We actually expect to have a pathological case like this, because we know that a WSS stochastic process with a spectral density given by (3.53) is not a "well-defined" one. If the line is is finite, and terminated with a finite resistance, then the spectral density of the process (which represents a voltage, since the transfer function is an impedance and the input is a current) at the output will have a lowest frequency below which it is constant, i.e. white [21], which then corresponds to a well-defined stochastic process with finite variance.

Another approach for a $1/f$ noise model is to use the summation of Lorentzian spectra [23]. This approach has been used in instrumentation to generate continuous-time $1/f$ noise over a specified range of frequencies. A sum of N Lorentzian spectra is given by

$$S(f) = \frac{2\sigma^2}{\pi} \sum_{h=1}^{N} \frac{\Phi_h}{\Phi_h^2 + f^2} \tag{3.56}$$

where Φ_h designate the pole-frequencies and f is the frequency [23]. It has been shown in [23] that $N = 20$ poles uniformly distributed over 14 decades are sufficient to generate $1/f$ noise over 10 decades with a maximum error less than 1%. Each Lorentzian spectrum in the summation in (3.56) can be easily obtained by using the thermal noise generator (which is modeled as a "constant modulated" version of the WSS Gaussian process as discussed before) of an LTI resistor R_h connected in parallel to a capacitance $C_h = C$, and their sum can be achieved by putting N of such $R_h - C_h$ groups in series [23], as shown in Figure 3.1. This is a MISO, stable, LTI system with inputs as the WSS Gaussian processes modeling the thermal noise current sources of the resistors.

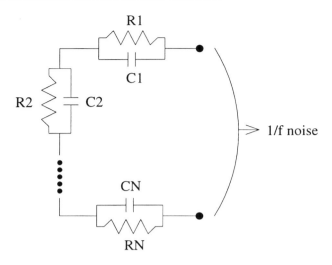

Figure 3.1. $1/f$ noise synthesizing circuit

Obviously, this system has internal states (represented by the capacitors), and hence it introduces new variables in the state vector of the system. Actually, this is inherent to $1/f$ noise, because the minimum amount of memory for a subsystem that exhibits $1/f$ noise is one state variable per decade of frequency [21].

The WSS process model with spectral density given by (3.53) was given for $1/f$ noise associated with a *time-invariant* bias. As for thermal and shot noise, we are interested in generalizing the $1/f$ noise model to the time-varying case. This was relatively straightforward for thermal and shot noise, since they are modeled as "delta-correlated" noise processes having independent values at every time point, and they were modeled as modulated WSS white Gaussian noise processes. In [1], it is pointed out that no $1/f$ noise is present in carbon resistors until a direct current (DC) is passed through the resistor [2]. Robinson in [18] mentions that any electronic device or component subjected to either a DC bias, or a *strong* AC (alternating current) signal can generate $1/f$ noise. Ambrozy in [24] reviews some experimental work investigating $1/f$ noise associated with DC bias and/or AC signals. van der Ziel in [25] mentions that if an AC current $I_o \cos(2\pi f_o t)$ is passed through a resistor, two $1/f$ noise *sidebands* appear around the frequency f_o.

In the next section, we will be discussing models for noise sources associated with a time-varying bias or signal, which can not be modeled as "delta-correlated" stochastic processes (and hence which can not be modeled as modulated WSS Gaussian processes). $1/f$ noise is certainly one such noise source, but our discussion in the next section will concentrate more generally on models for "correlated" noise, not just on $1/f$ noise. We will also discuss the observations we stated in the above paragraph on $1/f$ noise from several references in the context of the noise models we will investigate in the next section.

3.4.2 Models for correlated noise associated with time-varying signals

We assume that the "correlated" noise sources we are considering already have some kind of a WSS model for the time-invariant bias case. The WSS model for the time-invariant case could have either been derived from basic mechanisms that is behind this particular noise source, or it could have been derived from experimental investigation. For instance, the WSS model for $1/f$ noise has been derived from experimental investigation and it is characterized by the spectral density given by (3.53). We assume that the WSS model for the time-invariant case is characterized by a spectral density $S(f)$. We further assume that $S(f)$ is "separable" in the sense that it can be expressed as the product of two terms, one of which is a function of f but independent of bias quantities and hence sets the frequency dependence of $S(f)$, and the other term is a function of the bias quantities but is independent of f. Thus, we express $S(f)$ as

$$S(f) = m^2 \tilde{S}(f). \tag{3.57}$$

We assume that $m > 0$, and $\tilde{S}(f) > 0$ so that (3.57) is a sound spectral density. For instance, for $1/f$ noise modeled with (3.53), we can choose

$$m = \sqrt{(I^a)} \qquad \tilde{S}(f) = \frac{K}{f^b}. \tag{3.58}$$

We also assume that there exists a SISO, LTI system with transfer function $H(f)$ that satisfies

$$|H(f)|^2 = \tilde{S}(f). \tag{3.59}$$

When the bias is not time-invariant, i.e. $m(t)$ in (3.57) is a function of time, we will investigate the below cases as alternatives for the generalization of the WSS model described by (3.57):

1. The noise source is modeled as the output of a SISO, linear time-varying system which is a cascade of a memoryless modulator $m(t)$ (the output of which is basically the input multiplied with $m(t)$) and an LTI system with transfer function $H(f)$ that satisfies (3.59). The input to the system is the WSS white noise $\xi(t)$ with spectral density $S(f) = 1$.

2. The noise source is modeled as the output of a SISO, linear time-varying system which is a cascade of an LTI system with transfer function $H(f)$

that satisfies (3.59), and a memoryless modulator $m(t)$. The input to the system is the WSS white noise $\xi(t)$ with spectral density $S(f) = 1$. Note that this is *not equivalent* to the above model unless $m(t)$ is a constant function of time.

3. The noise source is modeled as a nonstationary process with time-varying spectral density given by
$$S(t, f) = m^2(t)\tilde{S}(f). \quad (3.60)$$

4. The noise source is modeled as a WSS process with spectral density given by
$$S(f) = \langle m^2 \rangle \tilde{S}(f), \quad (3.61)$$
where
$$\langle m^2 \rangle = \lim_{T \to \infty} \frac{1}{T} \int_{-T/2}^{T/2} m^2(t) dt, \quad (3.62)$$
whenever this limit exists.

5. The noise source is modeled as a WSS process with spectral density given by
$$S(f) = \langle m \rangle^2 \tilde{S}(f) \quad (3.63)$$
where
$$\langle m \rangle = \lim_{T \to \infty} \frac{1}{T} \int_{-T/2}^{T/2} m(t) dt \quad (3.64)$$
whenever this limit exists.

We will discuss the differences and similarities among these models, in terms of their probabilistic characterization, and also in terms of the implications of these models in their effect on a target system.

The first three models, in general for time-varying $m(t)$, are nonstationary. The last two are by definition WSS.

In the description of the models above, we have not made any assumptions on how $m(t)$ varies with time, so the models described above are general in that sense. For simplicity of discussion, and to be able to clearly see the similarities and differences of these models, for the rest of our discussion we will assume that $m(t)$ is a *periodically time-varying* function with period $T = 1/f_o$. And hence, we assume that $m(t)$ can be expanded into a Fourier series as follows
$$m(t) = \sum_{k=-\infty}^{\infty} b_k \exp\left(j 2\pi k f_o t\right). \quad (3.65)$$

Then $m^2(t)$ is also periodic with the same fundamental frequency f_o and we assume that it can also be expanded into a Fourier series as follows

$$m^2(t) = \sum_{k=-\infty}^{\infty} a_k \exp(j2\pi k f_o t). \tag{3.66}$$

Obviously the Fourier series coefficients a_k and b_k are related. In particular, they satisfy

$$a_0 = \sum_{k=-\infty}^{\infty} b_k b_{-k} = \sum_{k=-\infty}^{\infty} |b_k|^2, \tag{3.67}$$

which implies

$$a_0 \geq b_0^2. \tag{3.68}$$

Note that a_0 and b_0 are real for real $m(t)$. For periodic $m(t)$, we also have

$$a_0 = \langle m^2 \rangle \tag{3.69}$$

$$b_0 = \langle m \rangle \tag{3.70}$$

where $\langle m^2 \rangle$ and $\langle m \rangle$ are defined by (3.62) and (3.64).

3.4.2.1 Probabilistic characterization of the models. We would like to calculate the probabilistic characterization of the first model. Let us first consider the output Y of the memoryless modulator $m(t)$ with input $\xi(t)$. We can easily see that the autocorrelation of $Y(t) = m(t)\xi(t)$ is given by

$$R_Y(t,\tau) = \mathbf{E}\left[Y(t+\tau/2)Y(t-\tau/2)\right] = m^2(t)\delta(\tau) \tag{3.71}$$

since the autocorrelation of $\xi(t)$ is $R_\xi(\tau) = \delta(\tau)$. Then, the time-varying spectral density of Y is simply given by

$$S_Y(t,f) = m^2(t) = \sum_{k=-\infty}^{\infty} a_k \exp(j2\pi k f_o t) \tag{3.72}$$

which is independent of f. The Fourier series coefficients for $S_Y(t,f)$ are given by (see (2.52))

$$S_Y^{(k)}(f) = a_k. \tag{3.73}$$

Let us now consider the output Z of the LTI system with transfer function $H(f)$ that has Y as its input. The Fourier series coefficients for the time-varying spectral density $S_Z(t,f)$ of Z are then given by

$$\begin{aligned} S_Z^{(k)}(f) &= H(f+kf_o/2)H^*(f-kf_o/2)S_Y^{(k)}(f) \tag{3.74} \\ &= H(f+kf_o/2)H^*(f-kf_o/2)a_k \tag{3.75} \end{aligned}$$

which we obtained using (2.139), as a complete second-order probabilistic characterization for Z, which is a cyclostationary process and also the process that models the noise source.

Now, we consider the second model. Let us first consider the output Y of the LTI filter with transfer function $H(f)$ that has $\xi(t)$ as its input. Then, the spectral density of Y which is a WSS process is simply given by

$$S_Y(f) = |H(f)|^2 \tag{3.76}$$

and the autocorrelation of Y, by definition, is

$$R_Y(\tau) = \mathbf{F}^{-1}\{S_Y(f)\} = \mathbf{F}^{-1}\{|H(f)|^2\}. \tag{3.77}$$

Next, we consider the output Z of the memoryless modulator $m(t)$ with Y as its input. So, we have $Z(t) = m(t)Y(t)$. Then, the autocorrelation of $Z(t)$ can be expressed as

$$R_Z(t,\tau) = R_Y(\tau)[m(t+\tau/2)m(t-\tau/2)]. \tag{3.78}$$

If we substitute (3.65) in (3.78), we get

$$R_Z(t,\tau) = R_Y(\tau) \sum_{k=-\infty}^{\infty} \sum_{l=-\infty}^{\infty} b_k b_l \exp\left(j2\pi(k-l)f_o\tau/2\right) \exp\left(j2\pi(k+l)f_o t\right).$$

Now, we make a change of variables for the indexes of the double summation, $k + l = n$. After some rearrangement, we obtain

$$R_Z(t,\tau) = \sum_{n=-\infty}^{\infty} R_Y(\tau) \sum_{k=-\infty}^{\infty} b_k b_{n-k} \exp\left(j2\pi(2k-n)f_o\tau/2\right) \exp\left(j2\pi n f_o t\right). \tag{3.79}$$

In (3.79), we can identify the Fourier series coefficients of $R_Z(t,\tau)$ for Z (which is a cyclostationary process) as

$$R_Z^{(n)}(\tau) = R_Y(\tau) \sum_{k=-\infty}^{\infty} b_k b_{n-k} \exp\left(j2\pi(2k-n)f_o\tau/2\right). \tag{3.80}$$

Then, the Fourier series coefficients of the time-varying spectral density $S_Z(t,f)$ of Z are given by

$$\begin{aligned} S_Z^{(n)}(f) &= \mathbf{F}\left\{R_Y(\tau) \sum_{k=-\infty}^{\infty} b_k b_{n-k} \exp\left(j2\pi(2k-n)f_o\tau/2\right)\right\} \\ &= \mathbf{F}\{R_Y(\tau)\} \odot \mathbf{F}\left\{\sum_{k=-\infty}^{\infty} b_k b_{n-k} \exp\left(j2\pi(2k-n)f_o\tau/2\right)\right\} \end{aligned} \tag{3.81}$$

where we used (2.52). If we substitute (3.77) and

$$F\left\{\sum_{k=-\infty}^{\infty} b_k b_{n-k} \exp\left(j2\pi(2k-n)f_o\tau/2\right)\right\} = \sum_{k=-\infty}^{\infty} b_k b_{n-k}\delta(f - \frac{(2k-n)}{2}f_o)$$
(3.82)

in (3.81), we obtain

$$S_Z^{(n)}(f) = \sum_{k=-\infty}^{\infty} b_k b_{n-k} |H(f - \frac{(2k-n)}{2}f_o)|^2$$
(3.83)

which is a complete second-order probabilistic characterization of the cyclostationary process Z, which is the stochastic process that models the noise source.

For the third model, we already have an expression for the time-varying spectral density given by (3.60). If we substitute (3.66) and (3.59) in (3.60), we obtain

$$S(t,f) = m^2(t)\tilde{S}(f) = \sum_{k=-\infty}^{\infty} a_k |H(f)|^2 \exp(j2\pi k f_o t).$$
(3.84)

Now, we can easily identify the Fourier series coefficients of the time-varying spectral density $S(t, f)$ as

$$S^{(k)}(f) = a_k |H(f)|^2.$$
(3.85)

The fourth and fifth models are WSS with spectral densities given by (3.61) and (3.63) which fully characterize their second-order probabilistic characteristics. We rewrite (3.61) as

$$S(f) = a_0 |H(f)|^2.$$
(3.86)

and (3.63) as

$$S(f) = b_0 |H(f)|^2$$
(3.87)

3.4.2.2 Comparison of the models. Now, we discuss the differences and similarities among these five models characterized by (3.75), (3.83), (3.85), (3.86), and (3.87). Our first observation is obvious: For periodically time-varying $m(t)$, the first three models are cyclostationary processes, but the last two are WSS. If we compare the two WSS models considering (3.68), we observe that the fourth model describes a noise source with more total power than the one described by the fifth model.

Now, let us assume that $H(f) = \tilde{S}(f)$ satisfies

$$H(f) = 0 \quad \text{for} \quad f \geq \frac{f_o}{2}, \tag{3.88}$$

i.e. it is a *low pass* transfer function with a cut-off frequency which is less than $f_o/2$, the fundamental frequency of the periodic $m(t)$. If the noise source we are considering is $1/f$ noise, we can choose

$$H(f) = \sqrt{K}\frac{1}{\sqrt{f}}. \tag{3.89}$$

Obviously, this $H(f)$ does not have a cut-off frequency, but if f_o is "large enough", we will have

$$H(f) \approx 0 \quad \text{for} \quad f \geq \frac{f_o}{2}. \tag{3.90}$$

For $1/f$ noise, this is actually a reasonable approximation for many practical cases. We will further discuss this in the later chapters. If we evaluate the first model described by (3.75) considering (3.88), (3.75) reduces to

$$S_Z^{(k)}(f) = \begin{cases} a_0|H(f)|^2 & k = 0 \\ 0 & k \neq 0 \end{cases}. \tag{3.91}$$

Thus, with (3.88), we observe that the first model, which is cyclostationary in general, reduces to a WSS model, and moreover it becomes equivalent to the fourth model. This applies to $1/f$ noise model. Hence, the first and fourth model are in essence equivalent with (3.88) approximately satisfied.

Let us know consider that the noise sources modeled by the five models (characterized by (3.75), (3.83), (3.85), (3.86), and (3.87)) are inputs to a SISO, LTI system with transfer function $G(f)$. We assume that $G(f)$ satisfies

$$G(f) = 0 \quad \text{for} \quad f \geq \frac{f_o}{2} \tag{3.92}$$

i.e. it is a *low pass* transfer function with a cut-off frequency which is less than $f_o/2$, the fundamental frequency of the periodic $m(t)$. (It turns out that, for open-loop oscillators, a low-pass transformation which approximately satisfies the above condition can be defined between a noise source in the oscillator circuit and *phase noise*, a very important specification for oscillators. This fact, which we will later elaborate on extensively, is stated here to motivate for choosing such a system to compare the five noise models.) Now, we would like to calculate the probabilistic characterization of the output of the LTI system with the transfer function $G(f)$ when each of the five noise sources is the input to the system. With (3.92) and (3.88) satisfied, it can be shown that (using (2.139)) the output is a *WSS* process for all of the five noise models as inputs. Moreover, for the first, third and the fourth model, the spectral density of the output is given by

$$S(f) = a_0|H(f)|^2|G(f)|^2 \qquad (3.93)$$

and, for the second and the fifth model, the spectral density of the output is given by

$$S(f) = b_0^2|H(f)|^2|G(f)|^2. \qquad (3.94)$$

Recall, that $H(f)$ is the transfer function of a system that is part of the noise models, and $G(f)$ is the transfer function of the system the noise sources are affecting. We conclude that (with (3.92) and (3.88) satisfied), as far as we are concerned with the probabilistic characterization of the output of $G(f)$, the first, third and the fourth model are equivalent, and similarly the second and the fifth model are equivalent. From (3.68), the first, third and the fourth model predict a larger noise power at the output than the third and fifth do.

To test the validity or choose one of these noise models for a particular correlated noise source (e.g. $1/f$ noise), we would like to perform measurements. We will now assume that, one can measure the spectral density for a WSS noise source, and the zeroth order coefficient of the time-varying spectral density (which is basically the time-average) for a cyclostationary process. Let us compare the five noise models from this perspective, i.e. using the spectral density expressions for the WSS models, and only the zeroth order coefficient of the time-varying spectral density for the cyclostationary models, which we will denote by a subscript m_n for "measured" spectral density for the nth model.

$$S_{m_1} = a_0|H(f)|^2 \qquad (3.95)$$

$$S_{m_2} = \sum_{k=-\infty}^{\infty} b_k b_{-k}|H(f - kf_o)|^2 \qquad (3.96)$$

$$S_{m_3} = a_0|H(f)|^2 \qquad (3.97)$$

$$S_{m_4} = a_0|H(f)|^2 \qquad (3.98)$$

$$S_{m_5} = b_0^2|H(f)|^2 \qquad (3.99)$$

We observe that the "measured" spectral density of the first, third and the fourth model are exactly the same. The "measured" spectral density for the fifth model is also similar to these in terms of the shape of the frequency dependence. The only model that has frequency translated versions of $|H(f)|^2$ at all the harmonics of the fundamental frequency is the *second* model. van der Ziel in [25] mentions that if an AC current $I_o \cos(2\pi f_o t)$ is passed through a resistor, two $1/f$ noise *sidebands* appear around the frequency f_o. The only model that can exhibit this behavior is the second one. On the other hand, one has to be very careful in interpreting measurement results. Assuming that we are trying to measure the $1/f$ noise of a reasonably linear resistor biased with a periodic signal with fundamental frequency f_o, the $1/f$ noise sidebands at the harmonics of f_o could be arising from *nonlinearities* in our measurement set-up. Assuming that this is not the case, and if we observe $1/f$ noise sidebands at the harmonics of f_o, then we need to make our noise model compatible with this observation. Hence, the second model is the only choice. Of course, observing $1/f$ noise sidebands at the harmonics of f_o does not justify the second model completely. Even if we observe sidebands for the $1/f$ noise of a linear resistor of certain type, this does not imply that $1/f$ noise in all kinds of resistors, or other electronic components and devices will display the same behavior.

The main goal of our above discussion of alternatives for models for "correlated" noise sources was, most importantly, to pose the problem, and illustrate the properties of some simple and straightforward extensions of WSS models that have been used in practical noise calculations to the time-varying bias case, in a unified way using techniques from the *second-order* theory of stochastic processes. We hope that our discussion here will excite experimental and/or theoretical work on models for "correlated" noise sources that are associated with time-varying signals.

We would like to reemphasize the fact that even though the five noise models we have discussed turn out to be equivalent in special circumstances or seem to be equivalent when characterized in terms of "experimentally measurable" quantities (e.g. the zeroth order coefficient of the time-varying spectral density for a cyclostationary noise source), they are not equivalent in general. We believe that one can come up with practical electronic circuits, for which, using these "seemingly" equivalent noise models can yield completely different results in analyzing the noise performance.

3.5 Summary

In this chapter, we first reviewed the physical origins of *electrical* noise in electronic components and devices, concentrating on the most important ones, namely, *thermal*, *shot*, and $1/f$ noise. In particular, we discussed Nyquist's theorem on thermal noise in LTI dissipative systems. Then, we described stochastic process models for thermal and shot noise, which were generalizations of the WSS models associated with LTI resistors and time-invariant bias to LTV resistors and time-varying signals. The stochastic process models we described to model thermal and shot noise are, in general, nonstationary processes which are "modulated" versions of the standard WSS Gaussian white noise process, i.e. the formal derivative of the Wiener process. We have pointed out the problems in these generalizations, and the necessity of rigorously justifying them, especially the model for thermal noise associated with a nonlinear or time-varying resistor [22], for example the channel noise of a MOSFET transistor. We explained our motivation for requiring that all the noise source models be described in terms of the derivative of the Wiener process, so that the governing equations for the whole electronic circuit or system can be formulated as a system of stochastic differential equations. Then, we went on to discuss stochastic process models for "correlated" noise (e.g. $1/f$ noise) associated with time-varying signals. Our starting point was again the WSS models for time-invariant bias. We investigated several alternatives, calculated their second-order probabilistic characterization, and based on these characterizations analyzed several special cases, and pointed out the differences and similarities among these correlated noise models associated with time-varying signals.

Notes

1. The discussion of shot and thermal noise and their distinction is summarized from [17] and [18].

2. We would like to thank Ken Kundert for several fruitful discussions on this topic.

4 OVERVIEW OF NOISE SIMULATION FOR NONLINEAR ELECTRONIC CIRCUITS

In this chapter, we will present an overview of the techniques that have been proposed and used to analyze the effects of noise sources on the performance of electronic circuits and systems. We will discuss only general techniques that can be implemented for computer analysis. We do not claim to cover all the techniques that have been proposed in the literature. We especially exclude specialized ones that have been proposed for particular types of electronic circuits (e.g. noise analysis techniques for switched-capacitor circuits). We will go over the main characteristics of several different techniques, point out the similarities and differences among them, and briefly discuss their implementations for numerical computation.

4.1 Overview

In the previous chapter, we investigated the physical origins of electrical noise sources in electronic circuits or systems, and presented mathematical representations for them as stochastic processes. From this point on, when we refer to the noise sources in the system, we will be actually referring to their mathematical models as stochastic processes. To reach the final goal of simulating and characterizing the effect of these noise sources on the performance of an electronic circuit or system, the next thing we need is a mathematical representation of the system itself. As for many other systems, the dynamics of an electronic circuit can be described with a system of differential and algebraic equations. In this work, we will be mainly (except in Chapter 7) dealing with electronic circuits that can be modeled as an interconnection of basic network elements such as resistors, capacitors, inductors, controlled sources, independent voltage and current sources, etc. Resistors are usually described by possibly nonlinear relationships relating the terminal current to the terminal voltage. Similarly, controlled sources are described by possibly nonlinear relationships that relate the port voltages and currents. Capacitors and inductors are reactive elements, i.e. they can store energy, hence their descriptions involve time derivatives. For voltage-controlled capacitors, the terminal current is expressed as the time-derivative of the charge stored in the capacitor, and the charge is in general related to the terminal voltage with a nonlinear relationship. Similarly, for current-controlled inductors, the terminal voltage is expressed as the time-derivative of the flux of the inductor, and the flux is in general related to the terminal current with a nonlinear relationship. Resistors, capacitors and inductors are basic components of an electronic circuit. The models for semiconductor devices in electronic circuits are represented as an interconnection of the basic network elements. Given an interconnection of network elements, a system of mixed algebraic and differential equations can be written to describe the dynamics of the system. These equations consist of the Kirchoff's current law (KCL), the Kirchoff's voltage law (KVL) and the *branch* equations (i.e. the relationships that relate the terminal voltages and currents of the components). A particular way of formulating KVL, KCL and the branch equations, called the *Modified Nodal Analysis* (MNA) formulation, has been extensively used in circuit theory because of its generality and some other desirable properties for numerical computation. We will not further discuss the details of formulating network equations for electronic circuits, which is extensively covered elsewhere. For our purposes, it suffices to know that the governing equations that describe an electronic circuit can be formulated (using MNA) as a system of mixed algebraic and differential equations in the following form

$$\mathbf{I}(\mathbf{x},t) + \frac{d}{dt}\mathbf{Q}(\mathbf{x}) = 0 \qquad (4.1)$$

where $\mathbf{I}(\mathbf{x},t) : \mathbb{R}^n \times \mathbb{R} \to \mathbb{R}^n$ represents the "memoryless" network elements and the independent sources, and $\mathbf{Q}(\mathbf{x}) : \mathbb{R}^n \to \mathbb{R}^n$ represents the reactive elements, i.e the capacitors and the inductors. The state variables of the system, represented by the vector $\mathbf{x} \in \mathbb{R}^n$, consist of the *node voltages*, the *inductor currents* and the *independent voltage source currents*. There is one KCL equation in (4.1) for every node (except for the "ground" or reference node), which basically equates the sum of currents entering the node to the ones that are leaving it. The rest of the equations in (4.1) are the branch equations of inductors and voltage sources. The branch equations for the network elements other than the inductors and the voltage sources, and KVL equations are implicitly included in the MNA formulation.

Now that we have a mathematical representation of both the noise sources and the system, we can go ahead with the discussion of the noise analysis techniques. Note that, we include the deterministic excitations (i.e. the desired signals) in our mathematical description of the system (represented by the explicit time dependence of $\mathbf{I}(\mathbf{x}, t)$ in (4.1)). We assume that there is no "randomness" (e.g. random parameters) in the system itself. Even though the system is deterministic, when we have noise source excitations (modeled as stochastic processes) on the system, the state variables (e.g. node voltages) will be in general stochastic processes as opposed to deterministic signals. We formally define "noise analysis" or "noise simulation" of a system excited by noise sources to be the calculation of the *probabilistic characteristics* of the stochastic processes that represent the state variables. To make this definition precise we have to also define what we mean by "probabilistic characteristics". It could be a complete characterization of the fdds for the vector of stochastic processes that represent the state variables of the system, or for a second-order characterization, the autocorrelations/spectral densities as well as cross-correlations/cross-spectral densities of the stochastic processes that represent the state variables can be calculated. The several noise analysis techniques we are going to discuss can be classified in terms of what kind of a probabilistic characterization is being calculated. We will refer to the techniques that calculate spectral densities as *frequency-domain* techniques, and the techniques that calculate autocorrelation functions or moments as *time-domain* ones. We would like to point out that even though we refer to the techniques that calculate spectral densities as frequency-domain ones, this does not mean that the actual calculation is done entirely in frequency-domain. For instance, the time-varying spectral density for a nonstationary process could be calculated using a mixed frequency and time domain technique.

Frequency-domain techniques assume that the system is in some kind of *steady-state* condition, i.e. time-invariant, sinusoidal, periodic or quasi-periodic steady-state. Recall that we defined the system to include the deterministic excitations. In *forced* systems, the steady-state is set by the deterministic excitations on the system. In *autonomous* systems such as oscillators, the steady-state is set by the system itself. Moreover, frequency-domain techniques assume that the noise sources are WSS, cyclostationary or quasi-cyclostationary, as a natural consequence of the steady-state assumptions on the system itself. The steady-state properties of the noise sources are actually set by the system. For instance, as we have seen it in Chapter 3, shot noise is modeled as a modulated WSS Gaussian process, but the modulation is a function of the terminal current of the *pn* junction, which is set by the system.

For time-domain techniques, a steady-state condition on the system or the noise sources is not required. In general, one can have arbitrary time-varying deterministic excitations and hence nonstationary noise sources acting on the system. Thus, one can analyze *transient* noise phenomena with time-domain methods. In Chapter 6, we will see an example for how one can use the time-domain noise simulation technique we proposed to analyze a *seemingly* steady-state noise phenomenon, i.e. phase noise in free running oscillators, and arrive at some very useful results.

All but one (Monte Carlo noise simulation) of the noise simulation techniques we will be reviewing in this chapter treat the deterministic and noise excitations on the system *separately*. First, the nonlinear system is analyzed only with the deterministic excitations. The analysis with deterministic excitations is basically the solution of the system of differential equations given in (4.1). Then, using the results of this *large-signal* nonlinear analysis, a *linear* but possibly *time-varying* model for the system is constructed for noise analysis. The justification for such an approach lies in the assumption that noise signals are "small" signals and they do not "excite" the nonlinearities in the system. As a natural consequence of the separate treatment of the deterministic and noise excitations, the state variables (i.e. node voltages) are expressed as a summation of two terms, one of them due to the deterministic excitations and the other due to the noise sources. Each component is simulated or characterized separately. The small-signal analysis approach has been very popular in circuit theory. It is not only used for noise analysis but also for analyzing electronic circuits with large and small deterministic excitations present at the same time. In this case the nonlinear system is analyzed with the large excitations first, and then a linear model is constructed for the small excitations. In Chapter 5, we will formalize the small-signal approximation and linearization approach for noise sources modeled as stochastic processes.

4.2 Noise Simulation with LTI Transformations

In this method, a nonlinear circuit is assumed to have only *time-invariant* (DC) deterministic excitations, and it is also assumed that there exists a *time-invariant* steady-state solution. Note that, having only time-invariant excitations on the circuit does not necessarily imply that there exists a time-invariant steady-state solution. A time-invariant steady-state solution is obtained by setting the time derivative in (4.1) to zero and solving the resultant system of

OVERVIEW OF NOISE SIMULATION FOR NONLINEAR ELECTRONIC CIRCUITS   103

nonlinear algebraic equations, for instance, using Newton's method. This is referred to as the *DC analysis* of the circuit. It is possible to have more than one time-invariant steady-state solution. Then, the nonlinear circuit is *linearized* around the time-invariant steady-state solution to construct an LTI model for noise analysis. If this is applied to (4.1), we obtain

$$\mathbf{GX} + \mathbf{C}\frac{d}{dt}\mathbf{X} = 0 \qquad (4.2)$$

where $\mathbf{G}$ and $\mathbf{C}$ are $n \times n$ matrices, and the state vector $\mathbf{X} \in \mathbb{R}^n$, which is a vector of stochastic processes, represents the component of the node voltages and other circuit variables due to the noise sources. Obviously $\mathbf{X} = 0$ satisfies (4.2), because we have not included the noise sources in (4.2) yet. Let us connect a current noise source between two nodes in the LTI network represented by (4.2), which could be a noise source modeling the thermal noise of a resistor. (4.2) can be reformulated as below to include this noise source

$$\mathbf{GX} + \mathbf{C}\frac{d}{dt}\mathbf{X} + \mathbf{b}u(t) = 0, \qquad (4.3)$$

where $u(t)$ is the stochastic process that models the noise source, and $\mathbf{b}$ is an $n \times 1$ column vector with all entries set to zero except two, one of which is 1 and the other is -1. The two nonzero entries in $\mathbf{b}$ map the noise source to the nodes it is connected to, by modifying the corresponding KCL equations. The stochastic process $u(t)$ representing the noise source is assumed to be zero-mean, hence the current noise source does not have a direction and the 1 and -1 entries in $\mathbf{b}$ are interchangeable. Along with the time-invariant steady-state assumption, it is also assumed that both the noise sources, and the stochastic processes that are components of $\mathbf{X}$ (which model the components of the state variables due to noise) are WSS stochastic processes. Having only *WSS* noise sources does not necessarily imply that $\mathbf{X}$ is vector of *WSS* stochastic processes. A sufficient condition for this to be true is that the LTI system represented by the matrices $\mathbf{G}$ and $\mathbf{C}$ is *stable*.

We assume that we know the spectral density of the WSS noise source: $S_u(f)$. We would like to calculate the spectral densities of the WSS stochastic processes that are components of $\mathbf{X}$. We could accomplish this easily by using (2.131) if we can calculate the *transfer functions* from the noise source $u(t)$ to the components of $\mathbf{X}$. Let us now replace $u(t)$ in (4.3) with a complex exponential at frequency f, i.e. $\exp(j2\pi ft)$. Then, using the definition of the transfer function for an LTI system (given by (2.106)), we obtain

$$\mathbf{G}\mathbf{H}_b(f)\exp(j2\pi ft) + \mathbf{C}\frac{d}{dt}(\mathbf{H}_b(f)\exp(j2\pi ft)) + \mathbf{b}\exp(j2\pi ft) = 0 \quad (4.4)$$

where $\mathbf{H}_b(f)$ denotes the vector of transfer functions from the noise source to $\mathbf{X}$. After expanding the derivative in (4.4), we arrive at

$$\mathbf{GH}_b(f) + j2\pi f \mathbf{CH}_b(f) + \mathbf{b} = 0 \qquad (4.5)$$
$$(\mathbf{G} + j2\pi f \mathbf{C})\mathbf{H}_b(f) = -\mathbf{b}. \qquad (4.6)$$

Hence, we need to solve a linear system of equations given by (4.6) at every frequency point f to calculate the vector of transfer functions $\mathbf{H}_b(f)$. This could be accomplished with an LU decomposition of the matrix $\mathbf{T}(f) = \mathbf{G} + j2\pi f \mathbf{C}$, and a forward elimination and a backward substitution step. Then, the $n \times n$ matrix of spectral and cross-spectral densities for $\mathbf{X}$ is given by

$$\mathbf{S_X}(f) = \mathbf{H}_b(f) S_u(f) \mathbf{H}_b^T(f)^* \qquad (4.7)$$

where $S_u(f)$ is the spectral density of the noise source. (4.7) is a generalization of (2.131) (for a SISO LTI system) to the single-input multiple-output (SIMO) case.

In our above discussion, we considered a single noise source. One usually has a number of noise sources in the circuit. If we would like to calculate the spectral densities of the components of $\mathbf{X}$ due to all of the noise sources, we can repeat the above calculation for all of the noise sources to calculate a matrix of transfer functions from all of the noise sources to $\mathbf{X}$. Let us now assume that we have p noise sources (which could be correlated) with corresponding mapping vectors $\mathbf{b}_i$ $i = 1, 2, \ldots, p$. Then, the $n \times p$ matrix of transfer functions $\mathbf{H}(f)$ from all the noise sources to $\mathbf{X}$ is obtained as the solution of the system of equations

$$(\mathbf{G} + j2\pi f \mathbf{C})\mathbf{H}(f) = -[\mathbf{b}_1, \ldots, \mathbf{b}_p] = -\mathbf{B}. \qquad (4.8)$$

The ijth element of $\mathbf{H}(f)$ represents the transfer function from the jth noise source to the ith component of $\mathbf{X}$. (4.8) can be solved with a single LU decomposition of the matrix $\mathbf{T}(f) = \mathbf{G} + j2\pi f \mathbf{C}$ and p forward elimination and backward substitution steps at each frequency point f. It can be shown that the $n \times n$ matrix of spectral and cross-spectral densities for $\mathbf{X}$ is given by

$$\mathbf{S_X}(f) = \mathbf{H}(f) \mathbf{S}_u(f) \mathbf{H}^T(f)^* \qquad (4.9)$$

where the $p \times p$ matrix $\mathbf{S}_u(f)$ represents the matrix of spectral and cross-spectral densities for the noise sources. Most often, the noise sources are uncorrelated, hence $\mathbf{S}_u(f)$ is a diagonal matrix of the individual spectral densities of the noise sources. Even so, the matrix of spectral and cross-spectral densities for $\mathbf{X}$ is, in general, a full matrix. (4.9) is a generalization of (2.131) (for a SISO LTI system) to the MIMO case.

OVERVIEW OF NOISE SIMULATION FOR NONLINEAR ELECTRONIC CIRCUITS

One is usually interested in calculating the spectral density of a single output instead of the whole spectral, cross-spectral density matrix of $\mathbf{X}$. This output can usually be expressed as a linear combination of the components of $\mathbf{X}$ as follows:

$$Y(t) = \mathbf{d}^T \mathbf{X}(t) \quad (4.10)$$

where $\mathbf{d} \in \mathbb{R}^n$ is a constant vector. Most often, the output of interest is a voltage difference between two nodes. In this case, $\mathbf{d}$ has only two nonzero entries, set to 1 and -1, just like the $\mathbf{b}$ vectors that map the current noise sources to the nodes. Then, the spectral density of Y is given by

$$\begin{aligned} S_Y(f) &= \mathbf{d}^T \mathbf{S_X}(f) \mathbf{d} & (4.11) \\ &= \mathbf{d}^T \mathbf{H}(f) \mathbf{S}_u(f) \mathbf{H}^T(f)^* \mathbf{d} & (4.12) \end{aligned}$$

To calculate $S_Y(f)$, we need to calculate only the vector $\mathbf{d}^T \mathbf{H}(f)$ instead of the whole matrix $\mathbf{H}(f)$. From (4.8), we have

$$\mathbf{H}(f) = -\mathbf{T}(\mathbf{f})^{-1} \mathbf{B} \quad (4.13)$$

and hence

$$\begin{aligned} \mathbf{d}^T \mathbf{H}(f) &= (-\mathbf{d}^T \mathbf{T}(f)^{-1}) \mathbf{B} & (4.14) \\ &= \mathbf{w}(f)^T \mathbf{B} & (4.15) \end{aligned}$$

where the vector $\mathbf{w}(f)$ is the solution of the equation

$$\mathbf{T}(f)^T \mathbf{w}(f) = -\mathbf{d}. \quad (4.16)$$

Thus, to calculate $\mathbf{d}^T \mathbf{H}(f)$, we need a single LU decomposition of the matrix $\mathbf{T}(f)$ and a single forward elimination and backward substitution step at every frequency point f to solve the above equation with the right hand side set to $-\mathbf{d}$. The calculation of $\mathbf{d}^T \mathbf{H}(f)$ using (4.15) requires only n subtractions since each column of $\mathbf{B}$ has only two nonzero entries set to 1 and -1. The linear system of equations in (4.16) is called the *adjoint system*, because the LTI network represented by the matrix $\mathbf{T}(f)^T$ is the *adjoint* of the one that is represented by the untransposed matrix $\mathbf{T}(f)$. The idea of using the adjoint network concept in efficiently calculating the spectral density of a single output due to many noise sources was first proposed in [26]. Historically, the adjoint network method was derived through the use of Tellegen's theorem. This derivation is rather complicated compared with the pure matrix algebra approach we have presented above [27]. The adjoint network method is also used in efficiently performing sensitivity analysis of LTI networks.

Typically, one is interested in calculating the spectral density of the output, $S_Y(f)$, for a range of frequencies. Obviously, the frequency range of interest needs to be discretized for numerical computation. Then, the calculations described above are repeated for all of the frequency points. If we use a direct method (i.e. LU decomposition followed by elimination and substitution steps) to solve the linear system of equations in (4.16), then the cost of numerical computation is the same for all of the frequency points. To calculate the total power (i.e. $\mathbf{E}\left[Y(t)^2\right]$) of the output with this method, we would have to calculate the spectral density for the *whole* range of the frequencies where $S_Y(f)$ is not negligible and use a summation version of the integral in (2.47).

The noise analysis technique we described in this section is usually referred to as *AC noise analysis*, since it models a nonlinear circuit with an LTI network for noise analysis. It is available in almost every circuit simulator including SPICE [28], and it is widely used by analog circuit designers. The major limitation of this technique lies in the assumption that the nonlinear circuit has only time-invariant deterministic excitations and a time-invariant steady-state solution. Of course, all practical electronic circuits are supposed to deal with time-varying excitations. However, for certain types of circuits, such as amplifiers with excitations small enough so that they can be modeled as LTI networks for all purposes, this noise analysis method works very well. It is easy to implement, and it efficiently calculates the spectral density of the output for a range of frequencies. It is also easy to understand conceptually, because it is based on LTI network analysis and WSS stochastic processes. Unfortunately, it is not appropriate for noise analysis of circuits with "large" time-varying deterministic excitations or time-varying steady-state.

4.3 Noise Simulation with LPTV Transformations

In this method, a nonlinear electronic circuit is assumed to have only *periodically time-varying* (DC) deterministic excitations, and it is also assumed that there exists a *periodic* steady-state solution. (The methods to be presented can be easily generalized to circuits with a quasi-periodic steady-state, but to keep the discussion simple we will concentrate on the periodic case only.) Having only periodically time-varying excitations does not necessarily imply that there exists a periodic steady-state solution. The periodic steady-state solution for (4.1) can be obtained using several different techniques [29]: direct time-domain numerical integration (i.e. transient analysis), shooting method analysis, harmonic balance analysis, etc. Then, the nonlinear circuit is *linearized* around the periodic steady-state solution to construct an LPTV model for noise analysis. If this is applied to (4.1) we obtain

$$\mathbf{G}(t)\mathbf{X} + \mathbf{C}(t)\frac{d}{dt}\mathbf{X} = 0 \qquad (4.17)$$

where $\mathbf{G}(t)$ and $\mathbf{C}(t)$ are $n \times n$ periodically time-varying matrices as opposed to the constant ones in (4.2). Let us connect a current noise source between two nodes in the LPTV network represented by (4.17), which can then be reformulated as below to include the noise source

$$\mathbf{G}(t)\mathbf{X} + \mathbf{C}(t)\frac{d}{dt}\mathbf{X} + \mathbf{b}u(t) = 0 \qquad (4.18)$$

which is similar to (4.3) for the LTI case. Along with the periodic steady-state assumption, it is also assumed that both the noise sources and the stochastic processes that are components of $\mathbf{X}$ are cyclostationary stochastic processes. Note that, having only *cyclostationary* noise sources does not necessarily imply that $\mathbf{X}$ is vector of *cyclostationary* stochastic processes. A sufficient condition for this to be true is that the LPTV system represented by the matrices $\mathbf{G}(t)$ and $\mathbf{C}(t)$ is *stable*, i.e. all the Floquet exponents of the LPTV system should have negative real parts (see Section 2.4.9). As we will see in Chapter 6, this is not true for a free running oscillator.

We assume that we know the periodically time-varying spectral density of the cyclostationary noise source: $S_u(t, f)$. We would like to calculate the periodically time-varying spectral densities of the cyclostationary stochastic processes that are components of $\mathbf{X}$. To accomplish this, we need to calculate the *time-varying transfer functions* (see (2.110) for the definition) from the noise source $u(t)$ to the components of $\mathbf{X}$. Let us now replace $u(t)$ in (4.18) with a complex exponential at frequency f, i.e. $\exp(j2\pi ft)$. Then, using the definition of the time-varying transfer function for an LPTV system (see (2.111)), we obtain

$$\mathbf{G}(t)\mathbf{H}_b(f,t)\exp(j2\pi ft) + \mathbf{C}(t)\frac{d}{dt}(\mathbf{H}_b(f,t)\exp(j2\pi ft)) + \mathbf{b}\exp(j2\pi ft) = 0 \qquad (4.19)$$

where $\mathbf{H}_b(f,t)$ denotes the vector of periodically time-varying transfer functions from the noise source to $\mathbf{X}$. (4.19) is different from (4.4) in the sense that both the coefficient matrices $\mathbf{G}(t)$ and $\mathbf{C}(t)$ and the transfer function are *time-varying*.

Okumura et. al. in [30][1] propose to calculate a time discretized version of the periodically time-varying transfer function vector $\mathbf{H}_b(f,t)$. With this method, one places k time points in a period and approximate the time derivative in (4.19) using a numerical differentiation formula such as backward Euler, or more generally backward differentiation formula, etc. We will not present the details of this formulation. For details, the reader is referred to [30]. The resulting linear system of equations, the solution of which yields $\mathbf{H}_b(f,t)$ at k time points for a single frequency point f, is of dimension nk as opposed to the linear system of equations (4.6) of dimension n for the LTI case. A direct method (i.e. LU decomposition followed by forward elimination and backward substitution) is used to solve this system of equations in [30]. The large-signal analysis method that is used by [30] to calculate the periodic steady-state, and hence to construct the LPTV model for noise analysis, is the shooting method [29]. Telichevesky et. al. in [31][2] greatly improve the efficiency of the method proposed by [30]. [31] uses a matrix-implicit Krylov subspace based iterative method to solve the nk dimensional linear system for the discretized time-varying transfer function. [31] also makes use of a property of the Krylov subspace based iterative methods to efficiently calculate the discretized time-varying transfer function $\mathbf{H}_b(f,t)$ at different frequency points: Once the nk dimensional linear system of equations is solved for one frequency point f_1 using the matrix-implicit Krylov subspace based method, the numerical solution for another frequency point f_2 can reuse the results of some of the computations that were performed for f_1. In the above discussion, we have been considering the case where there is only one noise source. A straightforward generalization of the adjoint network formulation that was discussed for the LTI case can be applied to the LPTV case to calculate the time-varying transfer functions from many noise sources to a single output. Once the discretized transfer functions are calculated, one can calculate the spectral density of the output, which is assumed to be a cyclostationary process. Both [30] and [31] calculate *only* the zeroth order Fourier series coefficient (i.e. the time-average) of the periodically time-varying spectral density of the cyclostationary output (see (2.51)), i.e $S_Y^{(0)}(f)$, for a range of frequencies. This, obviously, is not a complete second-order probabilistic characterization of the output.

Hull et. al. [32, 33] propose a different method to calculate the periodically time-varying transfer functions. They first calculate the impulse responses for the LPTV transformations from the noise sources to the output, by solving a number of linear periodically time-varying systems of ordinary differential equations. The large-signal analysis with the deterministic excitations to calculate the periodic steady-state is performed with time domain numerical integration, i.e. transient analysis. Then, they use two-dimensional FFTs to calculate the Fourier series coefficients of the periodically time varying transfer functions (see (2.118)). Then, [33] calculates *only* the zeroth order Fourier series coefficient of the periodically time-varying spectral density of the cyclostationary output noise, but Hull [33] justifies this for his particular application, i.e. mixer noise analysis. He states that the mixers are almost always followed by an IF filter (a bandpass filter with bandwidth much smaller than the local oscillator frequency), and even if the noise at the output of the mixer is cyclostationary, the noise at the output of the IF filter is WSS.

Roychowdhury et.al. [34] present a frequency-domain method which can calculate a *complete* second-order probabilistic characterization of the cyclostationary output, i.e. they calculate not only the zeroth order Fourier series coefficient of the periodically time-varying spectral density, but also the higher order harmonics. They present a simple example with which they demonstrate the necessity of a full cyclostationary characterization as opposed to a *time averaged* characterization of the output. The frequency domain formulation presented by Roychowdhury [34] that uses the adjoint network concept yields in an efficient computational method to calculate a number of Fourier series coefficients of the time varying spectral density. The periodic steady-state with the large-signal deterministic excitations is calculated using the harmonic balance method [35], and the LPTV system for noise analysis is obtained by linearization. The Fourier series coefficients of the periodically time-varying transfer functions (see (2.118)) from the noise sources to the output for the LPTV system are calculated by solving a linear system of equations. In this case, this system is nh dimensional, where h is the number of harmonics used to represent the periodic steady-state. The efficient computation of the cyclostationary characterization of the output is obtained through factored-matrix methods together with preconditioned iterative techniques to solve the nh dimensional linear system. The authors also propose to use the PVL (Pade via Lanczos) [36] method and the shifted-QMR method to efficiently calculate the Fourier series coefficients of the time varying spectral densities for a range of frequencies.

4.4 Monte Carlo Noise Simulation with Direct Numerical Integration

The time domain Monte Carlo method is the only method, we discuss in this chapter, which does *not* treat the deterministic excitations and the noise signals separately. A linear model is *not* used for noise analysis. Instead, a number of time domain analyses are performed to simulate the full nonlinear model of the circuit under the influence of both the deterministic large excitations and the noise signals. The *nonlinear* circuit equations can be formulated (using MNA) including all the deterministic excitations and the noise sources to yield

$$\mathbf{I}(\mathbf{X},t) + \frac{d}{dt}\mathbf{Q}(\mathbf{X}) + \mathbf{B}(\mathbf{X})\mathbf{U}(t) = 0 \quad (4.20)$$

where $\mathbf{U}$ is a vector of stochastic processes representing the noise sources in the circuit, and $\mathbf{X}$ is a vector of stochastic processes (with a nonzero mean due to the deterministic excitations) representing the state variables, e.g. node voltages. We will discuss this formulation in detail in Chapter 5. With the time domain Monte Carlo technique, (4.20) is numerically integrated *directly* in time domain to generate a number of *sample paths* (at discrete time points) for the vector of stochastic processes $\mathbf{X}$ (see Section 2.5.7 and Section 2.2.12 for a discussion of direct numerical integration of stochastic differential equations and simulation of stochastic processes). Thus, an *ensemble* of sample paths is created. Then, by calculating various expectations over this ensemble, one can evaluate various probabilistic characteristics, including correlation functions and spectral densities. If one can prove that the vector of stochastic processes satisfies some ergodicity properties (see Section 2.2.11), it may be possible to calculate time averages over a *single* sample path to evaluate some time-averaged probabilistic characteristics which provide adequate information in some applications. This method is referred to as a *Monte Carlo* method, because in generating the sample paths using numerical integration, one has to realize or simulate the noise sources on the computer using a random number generator.

[37] uses a sum of sinusoids (with random phases) representation to realize the noise sources for time domain integration. [38] uses random amplitude pulse waveforms. Both [37] and [38] use standard numerical integration techniques (i.e. the transient analysis in circuit simulators) to generate sample paths for the node voltages. "Many" transient analyses of the circuit are performed with different sample paths of the noise sources.

The methods used by [37] and [38] have several problems. Shot and thermal noise sources in electronic circuits are ideally modeled as "white" noise sources. To simulate "white" noise sources accurately, one must either include very high frequency components in the sum of sinusoids representation, or set the pulse width to a very small value (to the correlation width of the noise sources, which is approximately 0.17 picoseconds for thermal noise at room temperature [5]) in the random amplitude pulse waveform representation. This limits the maximum time-step in numerical integration of (4.20) to a very small value making the simulation highly inefficient, which has to be repeated "many" times in a Monte Carlo fashion.

The portion due to noise sources in a waveform (i.e. sample path) obtained with a single run of the transient analysis will be much "smaller" (for most practical problems) when compared with the portion due to the deterministic signals. As a result, the sample paths obtained for different realizations of the noise sources will be very close to each other. These sample paths are only numerical approximations to the actual waveforms, therefore they contain *numerical noise* generated by the numerical integration algorithm. For instance, the variance of a node voltage as a function of time is calculated by taking averages over an ensemble of these sample paths. First, the mean as a function of time is calculated. Then, the variance at every time point is calculated by computing the average of the differences between the values from all of the sample paths and the mean. Consequently, the variance calculated for the node voltages includes the "noise" generated by the numerical algorithms. This creates another constraint on the time steps to be used during numerical integration.

Pseudo-random number generators on a computer often do not generate a large sequence of independent numbers, but reuse old random numbers instead. This can also become a problem if a circuit with many noise sources is simulated. This is usually the case, because every device has several noise sources associated with its model.

On the other hand, this method has a key property which might be very desirable in some applications: It does not assume that the noise signals are small compared with the deterministic excitations. The justification for using a linear model of the circuit for noise analysis is based on this assumption for all of the other methods we discuss in this chapter. For instance, the Monte Carlo technique can be useful in evaluating the noise performance of low noise amplifiers (LNAs) (a key component of the analog RF front-end of any wireless communication system). For LNAs, the desired deterministic signal can be weak enough so that its magnitude is comparable with the noise sources in the amplifier. In this case, the presence of other undesired large signals can preclude the use of an LTI model for AC small-signal and noise analysis.

4.5 Summary

We presented an overview of several techniques for noise analysis of nonlinear electronic circuits. We defined "noise analysis" to be the calculation of the probabilistic characteristics of the state variables (e.g. node voltages) of a circuit. The traditional noise analysis method that is based on LTI analysis and WSS noise sources assumes that the nonlinear circuit is in time-invariant steady-state under the influence of the deterministic excitations. With this method, the spectral density of the output is calculated at discrete frequency points in a frequency range of interest. Assuming that the output is a WSS process, this method calculates a complete second-order probabilistic characterization. We pointed out that the time-invariant steady-state assumption is not justified for many applications.

Then, we reviewed frequency domain techniques that are based on LPTV analysis with cyclostationary noise sources, which assume that the nonlinear circuit is in (quasi-) periodic steady-state under the influence of the deterministic excitations. In particular, a formulation based on the shooting method was proposed in [30]. Assuming that the output is also cyclostationary, this method calculates only a time-averaged probabilistic characterization. The efficiency of the method proposed in [30] was greatly improved by [31] using matrix-implicit Krylov subspace based iterative techniques for the solution of the large linear system of equations that needs to be solved to calculate the time discretized periodically time-varying transfer functions. The frequency domain steady-state technique proposed in [34] is based on a harmonic balance/conversion matrix formulation, and calculates a complete second-order characteristics of the cyclostationary output efficiently using factored matrix, preconditioned iterative techniques for the solution of the large linear system of equations that needs to be solved to calculate the frequency discretized periodically time-varying transfer functions.

We reviewed the time domain Monte Carlo noise analysis. This method is based on a direct numerical integration of the nonlinear differential equations that represent the circuit with the noise excitations to generate an ensemble of sample paths for the stochastic processes that represent the state variables. The information that is needed for a complete second-order characterization of the nonstationary output is contained in the ensemble of the sample paths generated with this method, even though a post-processing step is required to calculate spectral densities or correlation functions. We pointed out the efficiency and accuracy problems associated with this method. We believe that the results of rigorous systematic work on stochastic direct numerical simulation methods for differential equations involving stochastic processes have applications in electronic circuit and system design. Direct numerical simulation of the stochastic processes and the differential equations involving them can be useful in problems which can not be dealt with some simplifying approximations (i.e. linearization) that allow us to use the other noise analysis techniques reviewed in this chapter.

The time domain non-Monte Carlo method we proposed directly calculates the autocorrelation, cross-correlation matrix of the state variables of the system that are represented by nonstationary stochastic processes. With this method, one can analyze transient and nonstationary noise phenomena since a steady-state is not required. We discuss this method in the next chapter.

We believe that the combined use of frequency domain steady state techniques that calculate spectral densities and the time domain technique that calculates correlation functions will be most effective in investigating various important noise phenomena in electronic circuit design where the assumption that the effect of noise on the output is much smaller than the effect of desired deterministic signals (i.e. the linearization approximation) is justified. In problems where this assumption can not be justified, one has to revert to techniques, e.g. time domain Monte Carlo noise analysis, which treat the noise excitations and the deterministic signals together.

Notes

1. We believe that [30] has a formulation error regarding the linearization of nonlinear capacitors around the periodic steady-state.

2. The formulation presented in [31] is correct.

5 TIME-DOMAIN NON-MONTE CARLO NOISE SIMULATION

The time domain non-Monte Carlo (meaning that no random number generators are used) noise analysis technique we proposed [2, 3] is *not* restricted to circuits with a time-invariant or (quasi-) periodic steady-state with WSS or cyclostationary noise sources. The deterministic excitations on the circuit can be arbitrary time domain signals, including transient waveforms without a steady-state characteristics. As a result, the noise sources in the circuit will be *nonstationary* in general as opposed to being WSS or cyclostationary. All the circuit variables, i.e. node voltages, will also be nonstationary stochastic processes in general. A *complete* second-order probabilistic characterization would then require the calculation of the autocorrelation, cross-correlation matrix of the component of the state vector due to noise, which is given by

$$\mathbf{R}(t,\tau) = \mathbf{E}\left[\mathbf{X}(t+\tau/2)\mathbf{X}(t-\tau/2)^T\right] \quad (5.1)$$

or the time-varying spectral, cross-spectral density matrix

$$\mathbf{S_X}(t,f) = \mathbf{F}\left\{\mathbf{R}(t,\tau)\right\}. \quad (5.2)$$

In the general case, $\mathbf{R}(t,\tau)$ or $\mathbf{S}(t,f)$ are arbitrarily time varying functions of t, as opposed to a periodically time varying one. In this method, we directly calculate a time discretized version of $\mathbf{R}(t,\tau)$ numerically. For $\tau = 0$, the diagonal elements of $\mathbf{R}(t,\tau)$ gives us the noise variance (i.e. the total noise power) of the node voltages as a function of time, which can be directly calculated using this time domain non-Monte Carlo noise simulation technique.

Similar to the frequency domain methods, first, a large signal analysis of the circuit is performed with only the deterministic excitations. Then an LTV (as opposed to an LPTV one) model is constructed for noise analysis. Since this method does not require a large signal steady-state, the noise analysis can be performed along with the large signal analysis. This is useful in the sense that one does not need to store information about the large signal behavior of the circuit.

Apart from being able to analyze transient noise phenomena, this method is useful in applications where *total noise power* as a function of time is required. This is a natural output of this technique, which would be quite difficult to calculate using the frequency domain methods that calculate spectral densities.

In Chapter 6, we will present how one can use this technique to *investigate* and *model* an inherently nonstationary noise phenomenon, the so-called *phase noise* or *timing jitter* of oscillators, which is an extremely important concern in the design of clock generators for all kinds of applications and frequency synthesizers that generate the local oscillator (LO) signal in analog RF front-end of wireless communication systems.

We now present a detailed devlopment of the time domain non-Monte Carlo noise analysis algorithm for nonlinear electronic circuits, and describe its implementation for numerical computation. We also present noise simulation examples to illustrate the use of the noise simulation technique.

5.1 Formulation of Circuit Equations with Noise

The MNA formulation [39][27] of the mixed algebraic and differential equations that govern the behavior of a nonlinear electronic circuit including the deterministic excitations can be expressed as

$$\mathbf{I}(\mathbf{x},t) + \frac{d}{dt}\mathbf{Q}(\mathbf{x}) = 0 \tag{5.3}$$

where $\mathbf{I}(\mathbf{x},t) : \mathbb{R}^n \times \mathbb{R} \to \mathbb{R}^n$ and $\mathbf{Q}(\mathbf{x}) : \mathbb{R}^n \to \mathbb{R}^n$. The components of the vector $\mathbf{x} \in \mathbb{R}^n$ are the *state variables* of the nonlinear circuit. (See Section 4.1 for an explanation of this formulation.)

Under some rather mild conditions (which are satisfied by "well-modeled" circuits) on the continuity and differentiability of $\mathbf{I}$ and $\mathbf{Q}$, it can be proven that there exists a unique solution to (5.3) assuming that a fixed initial value $\mathbf{x}(0) = \mathbf{x}_0$ is given [39]. Let $\mathbf{x}_s$ be the deterministic solution to (5.3). Transient analysis in circuit simulators solves for $\mathbf{x}_s$ in time domain using numerical integration techniques for ordinary differential equations (ODEs) [39]. The initial value vector is obtained by a DC or operating point analysis (i.e. the solution of the system of nonlinear algebraic equations obtained from (5.3) by setting the time derivative to zero and using the $t = 0$ values for the independent sources) of the circuit before the numerical integration is started. For a nonlinear circuit, there may be more than one solution to the DC analysis problem.

We would like to include the noise sources in the formulation of the circuit equations in (5.3). Now, we state a number of assumptions on the models of the noise sources:

Assumption 5.1 *The noise sources in the circuit are modeled in terms of the standard white Gaussian noise process, i.e. the formal derivative of a standard Wiener process.*

Assumption 5.2 *If the model of a noise source contains a subsystem that has internal states, the variables that represent these internal states are included in the state vector* **x** *for the nonlinear circuit, and the equations that describe the noise modeling subsystem are appended to (5.3).*

For instance, an LTI low-pass subsystem can be used to filter a white Gaussian noise source, and the output of this subsystem is then used to model a WSS noise source with a low-pass spectral density.

Assumption 5.3 *All of the noise sources are modeled as current and voltage sources.*

These current and voltage sources are inserted into the model of an electronic device or component which is composed of an interconnection of basic network elements such as linear/nonlinear controlled sources, linear/nonlinear capacitors, linear/nonlinear resistors, etc. In practice, the noise sources of electronic components and devices are modeled as *current* sources, because current sources are more convenient to use with the MNA formulation.

Assumption 5.4 *The properties of the stochastic processes that represent the noise sources can be expressed in terms of the terminal voltages and currents of the device (or in terms of quantities that can be calculated using the terminal voltages and currents) and some fixed parameters such as the electron charge, the temperature and the Boltzmann's constant.*

All of the stochastic process models we have presented in Chapter 3 for thermal, shot and $1/f$ noise satisfy these assumptions. Any other model for a noise source that satisfies the above assumptions can be included in the formulation we are going to present.

With the above assumptions on the noise source models, we can modify the "deterministic" model of the system in (5.3) to formulate a system of *stochastic* algebraic and differential equations that describe the dynamics of the nonlinear circuit including the noise sources:

$$\mathbf{I}(\mathbf{X},t) + \frac{d}{dt}\mathbf{Q}(\mathbf{X}) + \mathbf{B}(\mathbf{X},t)\xi(t) = \mathbf{0}. \qquad (5.4)$$

We will refer to this formulation as the *NLSDE* (nonlinear stochastic differential equation) formulation. The last term in (5.4), $\mathbf{B}(\mathbf{x},t)\xi(t)$, that describes the noise sources, appears as an *additive* noise term. This is due to the assumptions we have stated above on the noise source models.

$\mathbf{X} \in \mathbb{R}^n$ in (5.4) is now a vector of stochastic processes that represents the state variables (e.g. node voltages) of the circuit. $\xi(t)$ in (5.4) is a p-dimensional vector of WSS white Gaussian processes (i.e. the formal derivative of the standard Wiener process).

Assumption 5.5 *The components of $\xi(t)$ are uncorrelated stochastic processes.*

This assumption is not required for the rest of our development, but it simplifies the discussion. Various noise sources in electronic devices usually have independent physical origin, and hence they are most often modeled as uncorrelated stochastic processes. For instance, all of the noise source models of semiconductor devices that are implemented in SPICE are uncorrelated. The generalization of the time domain noise analysis algorithm to handle correlated noise sources is straightforward, apart from slightly complicating the notation.

$\mathbf{B}(\mathbf{x},t)$ in (5.4) is an $n \times p$ matrix, the entries of which are functions of the state $\mathbf{x}$ and possibly t. In other words, $\mathbf{B}(\mathbf{x},t)$ is state and time dependent *modulation* for the vector of stochastic processes (or noise sources) $\xi(t)$. Every column of $\mathbf{B}(\mathbf{x},t)$ corresponds to a noise source in $\xi(t)$, and has normally either one or two nonzero entries. The rows of $\mathbf{B}(\mathbf{x},t)$ correspond to either a node equation (KCL) or a branch equation of an inductor or a voltage source. If there are no voltage noise sources in the circuit, there are no nonzero entries in the rows which correspond to the branch equations. Thus, $\mathbf{B}(\mathbf{x},t)$ *maps* the noise sources in $\xi(t)$ to the nodes they are connected to. Some semiconductor device models might contain two noise sources that are modeled with two *fully* correlated (which means that the noise sources are stochastically equivalent) stochastic processes. This situation can be handled naturally with this formulation by *mapping* the same noise source to different locations in the circuit. Thus, one can have two stochastically equivalent noise sources at different locations. Moreover, the modulations for these noise sources placed in $\mathbf{B}(\mathbf{x},t)$ need not be equal.

Thermal and shot noise sources as *modulated* white Gaussian noise sources are naturally included in this formulation. The modulation expressions to be placed in $\mathbf{B}(\mathbf{x}, t)$ for thermal and shot noise were described in Chapter 3.

For a current noise source that is connected between two nodes, the two nonzero entries (modulations) to be placed in $\mathbf{B}(\mathbf{x}, t)$ have equal absolute values but opposite signs. The sign of the modulations for the two nodes can be chosen arbitrarily, since a current noise source that is modeled with a zero mean stochastic process has no direction. For a voltage noise source there is only one nonzero entry in $\mathbf{B}(\mathbf{x}, t)$.

We will interpret (5.4) as an *Ito* system of stochastic differential equations. Now, we rewrite (5.4) in the more natural differential form

$$\mathbf{I}(\mathbf{X}, t)dt + d\mathbf{Q}(\mathbf{X}) + \mathbf{B}(\mathbf{X}, t)d\mathbf{W}(t) = 0 \qquad (5.5)$$

where we substituted $d\mathbf{W}(t) = \xi(t)dt$. $\mathbf{W}(t)$ is a vector of uncorrelated Wiener processes. Thus, our formulation is now in terms of the well-defined Wiener process, instead of the white Gaussian noise which does not exist in a strict mathematical sense.

5.2 Probabilistic Characterization of the Circuit with Noise

We formulated the equations that govern the behavior of a nonlinear circuit as a system of mixed stochastic algebraic and Ito stochastic differential equations. Now, we would like to "solve" (in some sense) this system of equations to generate "information" about the noise performance of the nonlinear circuit, which we refer to as the *noise analysis* of the circuit. In particular, we would like to calculate a probabilistic characterization of the vector of stochastic processes $\mathbf{X}(t)$ that represents the state variables of the system. The *finite-dimensional distributions* (i.e. fdds, see Section 2.2) for $\mathbf{X}(t)$ form a complete probabilistic characterization. If we can calculate the *joint probability density* for the state variables of the system, we can use the expectation operator over the calculated joint probability density to calculate all kinds of probabilistic characteristics of the state variables. In particular, we can obtain autocorrelation functions and spectral densities which can be calculated through second-order moments. It is also possible to calculate higher-order moments and hence a higher-order probabilistic characterization using the information contained in the joint probability density function of the state $\mathbf{X}(t)$. The joint probability density is a time dependent function $p(\mathbf{x}, t) : \mathbb{R}^n \times \mathbb{R} \to [0, \infty)$, because the stochastic processes that are components of $\mathbf{X}(t)$ are in general nonstationary. We have seen it in Section 2.5.6 that, given a system of Ito stochastic differential equations, one can derive a corresponding system of partial differential equations, called the Fokker-Planck equation, the solution of which gives us the conditional joint probability density function $p(\mathbf{x}, t|\mathbf{x}_0, t_0)$ for the state $\mathbf{X}(t)$. A Fokker-Planck equation that corresponds to (5.5) for a nonlinear circuit can be derived. Solving this Fokker-Planck equation *analytically* for the time-varying joint probability density for a practical nonlinear circuit is obviously out of the question. A numerical solution approach to this partial differential equation is also infeasible due to the large dimension (i.e. number of state variables) of the problem. Obtaining analytical or numerical solutions for the Fokker-Planck equation for a nonlinear circuit is only possible for simple "toy" circuits with a state-space dimension that is smaller than 5.[1]

Remark 5.1 *Calculation of the conditional joint probability density function $p(\mathbf{x}, t|\mathbf{x}_0, t_0)$ for the state $\mathbf{X}(t)$ by solving the Fokker-Planck equation that corresponds to (5.5) for a practical nonlinear circuit is not feasible.*

Because of great difficulties in obtaining nonstationary solutions of the Fokker-Planck equation that corresponds to (5.5) for a nonlinear circuit, we could attempt to calculate the time-varying moments of $\mathbf{X}(t)$ separately, instead of calculating the whole joint probability density which, in a sense, is equivalent to calculating all orders of moments at the same time. To accomplish this, we could derive differential equations, the solution of which will give us the time-varying moments of $\mathbf{X}(t)$. The differential equations for moments can be obtained in two ways [13]. One way is to use the Fokker-Planck equation. However, another and more natural way of deriving equations for moments is based on applying Ito's formula (see Section 2.5.4) for stochastic differentials to the function

$$h(\mathbf{X}(t)) = X_1^{k_1}(t) X_2^{k_2}(t) \ldots X_n^{k_n}(t) \qquad (5.6)$$

(where $X_1(t), X_2(t), \ldots, X_n(t)$ are the stochastic processes that are components of $\mathbf{X}(t)$) to calculate an expression for $d/dt\, h(\mathbf{X}(t))$. Then, the expectation of this expression is taken to calculate $d/dt\, \mathrm{E}\left[h(\mathbf{X}(t))\right]$ and hence to derive the differential equations for the moments. This is very similar to what we did to derive the Fokker-Planck equation in Section 2.5.6 for the one-dimensional case. Here, we are using the specific $h(\mathbf{X}(t))$ in (5.6), which is specifically chosen to extract the moments. For a system of nonlinear Ito stochastic differential equations such as (5.5), the differential equation for a lower order moment may contain terms of higher order moments. Thus,

Remark 5.2 *For a system of nonlinear Ito stochastic differential equations such as (5.5), in general, an* infinite hierarchy *of moment equations [13] is obtained.*

In order to obtain a closed form of moment equations, some *closure* approximations have to be introduced to truncate the infinite hierarchy. In a practical problem, one is usually interested only in some lower order moments. Closure approximations have been proposed in the literature in solving stochastic problems in several disciplines such as turbulence theory, control and vibratory systems [13]. One of the simplest closure approximations that first comes to mind is the *Gaussian* closure scheme, according to which, higher order moments are expressed in terms of the first and second order moments as if the components of **X** are Gaussian stochastic processes. Recall that the distribution of a Gaussian random variable is completely characterized with its mean and variance, i.e. the first and second order moments. Variations on the basic Gaussian closure scheme have been proposed for different kinds of problems in the literature. Closure schemes are designed using specific characteristics of the problem in hand. Not all the closure approximations proposed in the literature have sound mathematical basis, but they can be justified using the physical properties of the specific problem they are being used for [13]. Using these closure schemes, one arrives at systems of *nonlinear* ordinary differential equations for various moments. By solving these equations numerically, one can calculate the time evolution of the moments of the state $\mathbf{X}(t)$.

We now point out some properties of the problem we are trying to solve. The system we are dealing with is a nonlinear electronic circuit with electrical (e.g. shot and thermal) noise sources. In general, the noise signals can be considered as small excitations on the system. In this case, the time development of the system will almost be deterministic, and the fluctuations due to noise will be small perturbations. This observation immediately brings to mind the technique of small-signal analysis that is widely used in the analysis and design of electronic circuits (and in many other problems), i.e. modeling the system with a *linear* (but possibly time-varying) model for noise signals.

Remark 5.3 *If the inputs to a stable linear (possibly time-varying) system are Gaussian stochastic processes (satisfied by shot and thermal noise), then the state* **X** *and the outputs of the system are also Gaussian, and in general nonstationary, stochastic processes.*

We know that a Gaussian process is completely characterized by the first and second-order moments. Thus,

Remark 5.4 *The knowledge of the time evolution of only the mean*

$$\mathrm{E}\left[\mathbf{X}(t)\right] \tag{5.7}$$

and the autocorrelation function

$$\mathrm{E}\left[\mathbf{X}(t+\tau/2)\mathbf{X}(t-\tau/2)^T\right] \tag{5.8}$$

is sufficient for a complete probabilistic characterization of a vector of Gaussian stochastic processes.

Remark 5.3 and Remark 5.4 are the key facts why the small-signal analysis technique greatly simplifies the noise analysis problem for a nonlinear system.

Treating the noise sources as small deterministic excitations and deriving a small-signal analysis scheme with such a consideration is not mathematically sound. The small deterministic signals satisfy certain smoothness conditions. For small-signal analysis, they are usually modeled with sinusoidal signals. On the other hand, noise signals in general do not satisfy smoothness conditions (i.e. continuity and differentiability) and usually exhibit erratic behavior. Thus, the derivation of a small-signal analysis scheme for noise signals should be done using approximation techniques developed for stochastic processes and systems. An approximate technique developed in the analysis of engineering nonlinear dynamical systems subjected to random excitations is the *perturbation* method [13]. The idea behind this method comes from the classical theory of ordinary differential equations. In this method, a small parameter $\epsilon \ll 1$ is usually introduced to the system, and the solution is expressed in the form of an expansion with respect to the powers of ϵ. In the next section, we will apply such a stochastic perturbation scheme to (5.5). The result we are going to obtain at the end of the next section can also be derived using deterministic arguments through the use of first-order Taylor's expansion of nonlinearities in (5.5) [3], but some of the steps in this derivation are not mathematically sound (which has to do with considering the "derivative" of a stochastic process using deterministic arguments).

5.3 Small Noise Expansion

We will apply a stochastic perturbation scheme [10] to the NLSDE formulation given in (5.5) for a nonlinear circuit with noise sources. We introduce a parameter ϵ to (5.5) as a multiplying factor in the last term that describes the noise sources. To keep (5.5) invariant, we then need to modify the entries of $\mathbf{B}(\mathbf{X}, t)$, i.e. divide them by ϵ. For the perturbation approximation to be justified, we need to have $\epsilon \ll 1$. We now state

Assumption 5.6 *If we set ϵ to a value such that the "magnitude" of $\mathbf{B}(\mathbf{X}, t)/\epsilon$ is "comparable" to the "magnitude" of the deterministic excitations (which are represented by the explicit t dependence of $\mathbf{I}(\mathbf{x}, t)$ in (5.5)), then we assume that $\epsilon \ll 1$ is satisfied for a practical nonlinear electronic circuit with electrical noise sources.*

Assumption 5.6 is the assumption that will be used to justify the perturbation approximation. For the notational simplicity of the discussion below, we will keep the entries of $\mathbf{B}(\mathbf{X}, t)$ unchanged, i.e. we will not divide them by ϵ, but we will introduce ϵ as a multiplying factor for the noise term in (5.5):

$$\mathbf{I}(\mathbf{X},t)dt + d\mathbf{Q}(\mathbf{X}) + \epsilon \mathbf{B}(\mathbf{X},t)d\mathbf{W}(t) = 0. \tag{5.9}$$

If we substitute $\epsilon = 1$ in (5.9), we obtain the original NLSDE formulation (5.5). If $\epsilon = 0$, then (5.5) reduces to (5.3) which describes the nonlinear circuit without the noise sources. The solution for (5.9) will be written as an expansion with respect to the powers of ϵ. Thus,

Assumption 5.7 *The solution* $\mathbf{X}(t)$ *of (5.9) can be written as*

$$\mathbf{X}(t) = \mathbf{X}_0(t) + \epsilon \mathbf{X}_1(t) + \epsilon^2 \mathbf{X}_2(t) + \ldots \tag{5.10}$$

where $\mathbf{X}_0(t), \mathbf{X}_1(t), \mathbf{X}_2(t), \ldots$ *are all n-dimensional vectors of stochastic processes.*

Assumption 5.8 $\mathbf{I}(\mathbf{x},t)$ *can expanded as*

$$\begin{aligned}\mathbf{I}(\mathbf{x},t) &= \mathbf{I}(\mathbf{x}_0 + \epsilon\mathbf{x}_1 + \epsilon^2\mathbf{x}_2 + \ldots, t) & (5.11)\\ &= \mathbf{I}_0(\mathbf{x}_0,t) + \epsilon\mathbf{I}_1(\mathbf{x}_0,\mathbf{x}_1,t) + \epsilon^2\mathbf{I}_2(\mathbf{x}_0,\mathbf{x}_1,\mathbf{x}_2,t) + \ldots & (5.12)\end{aligned}$$

where $\mathbf{I}_0(\mathbf{x}_0,t), \mathbf{I}_1(\mathbf{x}_0,\mathbf{x}_1,t), \ldots$ *are all vector-valued deterministic functions.*

Now, not to complicate the notation, we assume that both $\mathbf{x}$ and $\mathbf{I}$ are scalars to demonstrate how one can obtain the expansion in (5.12).

Lemma 5.1 *For scalar* $\mathbf{x}$ *and* $\mathbf{I}$*, we have*

$$\begin{aligned}I(x,t) &= I(x_0 + \sum_{i=1}^{\infty} \epsilon^i x_i, t) & (5.13)\\ &= \sum_{j=0}^{\infty} \frac{1}{j!} \frac{\partial^j I(x_0,t)}{\partial x_0^j} (\sum_{i=1}^{\infty} \epsilon^i x_i)^j. & (5.14)\end{aligned}$$

Proof. This result is easily obtained by using Taylor's expansion [10]. □

By expanding (5.14), we can calculate the zeroth and first order coefficients, $I_0(x_0,t)$ and $I_1(x_0,x_1,t)$, in (5.12) to be

$$I_0(x_0,t) = I(x_0,t) \tag{5.15}$$

$$I_1(x_0,x_1,t) = \frac{\partial I(x_0,t)}{\partial x_0} x_1. \tag{5.16}$$

(5.14) can be generalized to the multidimensional case when $\mathbf{x}$ and $\mathbf{I}$ are n-vectors. In this case, we obtain

$$\mathbf{I}_0(\mathbf{x}_0,t) = \mathbf{I}(\mathbf{x}_0,t) \tag{5.17}$$

$$\mathbf{I}_1(\mathbf{x}_0,\mathbf{x}_1,t) = \frac{\partial \mathbf{I}(\mathbf{x}_0,t)}{\partial \mathbf{x}_0} \mathbf{x}_1 \tag{5.18}$$

for the zeroth and first order coefficients in (5.12). We define

$$\mathbf{G}(\mathbf{x}_0, t) = \frac{\partial \mathbf{I}(\mathbf{x}_0, t)}{\partial \mathbf{x}_0} \tag{5.19}$$

which is an $n \times n$ matrix-valued function. Even though it is not easy to write explicitly the full set of coefficients in general for the expansion (5.12), it is easy to see the following:

Lemma 5.2 *For $i \geq 1$, one can write*

$$\mathbf{I}_i(\mathbf{x}_0, \mathbf{x}_1, \ldots, \mathbf{x}_i, t) = \mathbf{G}(\mathbf{x}_0, t)\mathbf{x}_i + \check{\mathbf{I}}_i(\mathbf{x}_0, \mathbf{x}_1, \ldots, \mathbf{x}_{i-1}, t) \tag{5.20}$$

where $\check{\mathbf{I}}_i$ is independent of $\mathbf{x}_i$.

Proof. See [10]. Note that $\check{\mathbf{I}}_1 = 0$. The validity of (5.20) can be easily seen by examining (5.14). □

Similar to Assumption 5.8, we state

Assumption 5.9 $\mathbf{Q}(\mathbf{x})$ *and* $\mathbf{B}(\mathbf{x}, t)$ *can be expanded as*

$$\begin{aligned}
\mathbf{Q}(\mathbf{x}) &= \mathbf{Q}_0(\mathbf{x}_0) + \epsilon \mathbf{Q}_1(\mathbf{x}_0, \mathbf{x}_1) + \epsilon^2 \mathbf{Q}_2(\mathbf{x}_0, \mathbf{x}_1, \mathbf{x}_2) + \ldots & (5.21) \\
\mathbf{B}(\mathbf{x}, t) &= \mathbf{B}_0(\mathbf{x}_0, t) + \epsilon \mathbf{B}_1(\mathbf{x}_0, \mathbf{x}_1, t) + \epsilon^2 \mathbf{B}_2(\mathbf{x}_0, \mathbf{x}_1, \mathbf{x}_2, t) + \ldots & (5.22)
\end{aligned}$$

where

$$\mathbf{Q}_0(\mathbf{x}_0) = \mathbf{Q}(\mathbf{x}_0) \tag{5.23}$$

$$\mathbf{Q}_1(\mathbf{x}_0, \mathbf{x}_1) = \frac{\partial \mathbf{Q}(\mathbf{x}_0)}{\partial \mathbf{x}_0} \mathbf{x}_1 \tag{5.24}$$

and

$$\mathbf{B}_0(\mathbf{x}_0, t) = \mathbf{B}(\mathbf{x}_0, t). \tag{5.25}$$

Recall that $\mathbf{B}(\mathbf{x}, t)$ is an $n \times p$ matrix-valued function, unlike $\mathbf{I}(\mathbf{x}, t)$ and $\mathbf{Q}(\mathbf{x})$ which are vector-valued functions. We define the $n \times n$ matrix-valued function

$$\mathbf{C}(\mathbf{x}_0) = \frac{d\mathbf{Q}(\mathbf{x}_0)}{d\mathbf{x}_0} \tag{5.26}$$

which appears in the expansion of $\mathbf{Q}(\mathbf{x})$ exactly in the same way $\mathbf{G}(\mathbf{x}_0, t)$ appears in the expansion of $\mathbf{I}(\mathbf{x}, t)$. Similar to Lemma 5.2, we have

Lemma 5.3 *For $i \geq 1$, one can write*

$$\mathbf{Q}_i(\mathbf{x}_0, \mathbf{x}_1, \ldots, \mathbf{x}_i) = \mathbf{C}(\mathbf{x}_0)\mathbf{x}_i + \check{\mathbf{Q}}_i(\mathbf{x}_0, \mathbf{x}_1, \ldots, \mathbf{x}_{i-1}) \tag{5.27}$$

where $\check{\mathbf{Q}}_i$ is independent of $\mathbf{x}_i$.

Note that $\check{\mathbf{Q}}_1 = \mathbf{0}$. Now, we are ready to state

Theorem 5.1 $\mathbf{X}_0(t), \mathbf{X}_1(t), \mathbf{X}_2(t), \ldots$ in (5.10) satisfy the following infinite set of systems of stochastic differential equations for $i \geq 1$:

$$\mathbf{I}(\mathbf{X_0}, t)dt + d\mathbf{Q}(\mathbf{X_0}) = 0 \tag{5.28}$$

$$\begin{aligned}&[\mathbf{G}(\mathbf{X}_0, t)\mathbf{X}_i + \check{\mathbf{I}}_i(\mathbf{X}_0, \mathbf{X}_1, \ldots, \mathbf{X}_{i-1}, t)]dt \\ &+ d[\mathbf{C}(\mathbf{X}_0)\mathbf{X}_i + \check{\mathbf{Q}}_i(\mathbf{X}_0, \mathbf{X}_1, \ldots, \mathbf{X}_{i-1})] \\ &+ \mathbf{B}_{i-1}(\mathbf{X}_0, \mathbf{X}_1, \ldots, \mathbf{X}_{i-1}, t)d\mathbf{W}(t) = 0.\end{aligned} \tag{5.29}$$

Proof. We substitute the expansions (5.10), (5.12), (5.21) and (5.22) along with (5.20) and (5.27) in the system of stochastic differential equations (5.9). We equate the coefficients of the like powers of ϵ. (5.28) is obtained by equating the terms that are independent of ϵ. By equating the coefficients of the ith powers of ϵ, we obtain (5.29). □

(5.28) and (5.29) can now be treated (i.e. solved) sequentially. (5.28) is exactly (5.3) which describes the nonlinear circuit *without* the noise sources, and hence has a solution $\mathbf{x}_s$ as defined in Section 5.1. Thus, we have

$$\mathbf{X}_0(t) = \mathbf{x}_s(t). \tag{5.30}$$

The solution $\mathbf{X}_0(t)$ is deterministic provided that the initial conditions that are specified for (5.28) are deterministic. If we set $i = 1$ in (5.29), we obtain

$$\mathbf{G}(\mathbf{x}_s, t)\mathbf{X}_1 dt + d[\mathbf{C}(\mathbf{x}_s)\mathbf{X}_1] + \mathbf{B}(\mathbf{x}_s, t)d\mathbf{W}(t) = 0 \tag{5.31}$$

where we used $\check{\mathbf{I}}_1 = \mathbf{0}$, $\check{\mathbf{Q}}_1 = \mathbf{0}$ and (5.30). Let us define

$$\begin{aligned}\mathbf{G}(t) &= \mathbf{G}(\mathbf{x}_s(t), t) \\ \mathbf{C}(t) &= \mathbf{C}(\mathbf{x}_s(t)) \\ \mathbf{B}(t) &= \mathbf{B}(\mathbf{x}_s(t), t)\end{aligned} \tag{5.32}$$

and rewrite (5.31) as

$$\mathbf{G}(t)\mathbf{X}_1 dt + d[\mathbf{C}(t)\mathbf{X}_1] + \mathbf{B}(t)d\mathbf{W}(t) = 0. \tag{5.33}$$

The solution of (5.31), $\mathbf{X}_1(t)$, is the first-order term in the expansion (5.10) and is called a multivariate time-varying *Ornstein-Uhlenbeck* process. For our purposes, the first-order term $\mathbf{X}_1(t)$ is quite adequate to characterize the solution of (5.9), and it amounts to a linearization of the original system (5.9) about the deterministic solution. We approximate the solution $\mathbf{X}(t)$ of (5.9) with

$$\mathbf{X}(t) \approx \mathbf{X}_0(t) + \epsilon \mathbf{X}_1(t). \tag{5.34}$$

To solve for $\mathbf{X}(t)$, we first calculate $\mathbf{X}_0(t) = \mathbf{x}_s(t)$ as the solution of (5.3), which is deterministic. Then, we use the calculated $\mathbf{x}_s(t)$ to calculate $\mathbf{G}(t)$, $\mathbf{C}(t)$ and $\mathbf{B}(t)$ (which are all deterministic) using (5.32), (5.19) and (5.26). $\mathbf{X}_1(t)$ is obtained as the solution of the system of *linear* stochastic differential equations (5.33) with *time-varying* coefficient matrices. Calculating higher-order terms in (5.10) is more complicated because of the more complex form of (5.29) for $i \geq 2$, but, in essence, they are treated in exactly the same way. In order to solve the equation for $\mathbf{X}_i(t)$, we assume we know all the $\mathbf{X}_j(t)$ for $j < i$ so that $\check{\mathbf{I}}_i$, $\check{\mathbf{Q}}_i$ and $\mathbf{B}_{i-1}$ become known *stochastic* functions of t after substituting these solutions. Then (5.29) becomes

$$[\mathbf{G}(t)\mathbf{X}_i + \check{\mathbf{I}}_i(t)]dt + d[\mathbf{C}(t)\mathbf{X}_i + \check{\mathbf{Q}}_i(t)] + \mathbf{B}_{i-1}(t)d\mathbf{W}(t) = 0 \tag{5.35}$$

where $\mathbf{G}(t)$ and $\mathbf{C}(t)$ are deterministic, and $\check{\mathbf{I}}_i(t)$, $\check{\mathbf{Q}}_i(t)$ and $\mathbf{B}_{i-1}(t)$ are stochastic functions of time.

5.4 Derivation of a Linear Time Varying SDE Model for Noise Analysis

We now concentrate on (5.33) which we repeat below with the substitution $\mathbf{X}_1(t) = \mathbf{X}_n(t)$, to denote the stochastic noise component in the expansion (5.34) for the total solution,

$$\mathbf{G}(t)\mathbf{X}_n dt + d[\mathbf{C}(t)\mathbf{X}_n] + \mathbf{B}(t)d\mathbf{W}(t) = 0. \tag{5.36}$$

The time-varying matrix $\mathbf{G}(t)$ represents the linear and nonlinear resistive components in the circuit linearized around the deterministic solution $\mathbf{x}_s(t)$, and the time-varying matrix $\mathbf{C}(t)$ represents the linear and nonlinear reactive components (i.e. capacitors and inductors) linearized around the deterministic solution $\mathbf{x}_s(t)$. We expand the second term above to obtain

$$\mathbf{G}(t)\mathbf{X}_n dt + [\frac{d\mathbf{C}(t)}{dt}]\mathbf{X}_n dt + \mathbf{C}(t)d\mathbf{X}_n + \mathbf{B}(t)d\mathbf{W}(t) = 0. \tag{5.37}$$

Notice that we used a classical calculus differentiation rule above. In this case, the Ito stochastic differentiation rule is same as the classical one, because the argument is merely *linear* in $\mathbf{X}_n$. The time-varying matrix $\dot{\mathbf{C}}(t) = d/dt\mathbf{C}(t)$ in (5.37) represents the time-derivative of the time-varying capacitors in the linearized circuit. With reorganization, (5.37) becomes

$$\mathbf{A}(t)\mathbf{X}_n dt + \mathbf{C}(t)d\mathbf{X}_n + \mathbf{B}(t)d\mathbf{W}(t) = 0 \tag{5.38}$$

where
$$\mathbf{A}(t) = \mathbf{G}(t) + \dot{\mathbf{C}}(t). \tag{5.39}$$

Theorem 5.2 *The system of stochastic differential equations in (5.38) (obtained from an MNA formulation) can be transformed into the standard state equations form*
$$d\mathbf{Y} = \mathbf{E}(t)\,\mathbf{Y} + \mathbf{F}(t)\,d\mathbf{W}(t) \tag{5.40}$$
where $\mathbf{Y}$ is an m-dimensional vector, $\mathbf{E}(t)$ is $m \times m$, $\mathbf{F}(t)$ is $m \times p$ with $m \leq n$, if the following conditions are satisfied:

- *There are no nodes (except for the ground node) in the circuit with a connection both to a capacitor and a voltage source (independent or dependent).*

- *Every node in the circuit with a connection to a capacitor has a* capacitive path *to ground or an independent voltage source node.*

Proof. $\mathbf{C}(t)$ is not a full-rank matrix in general, it may have zero rows and columns. For instance, if a circuit variable is a node voltage, and if this node does not have any capacitors connected to it in the circuit, then all of the entries in the column of $\mathbf{C}(t)$ corresponding to this circuit variable will be zero for all t. Also, the node equation (KCL) corresponding to this node will not contain any time-derivatives, hence the row of $\mathbf{C}(t)$ corresponding to this node equation will also be zero for all t. Thus, some of the rows and columns of $\mathbf{C}(t)$ are *structurally* zero, independent of t. Moreover, the number of zero rows is equal to the number of zero columns. If we reorder the variables in $\mathbf{X}_n$ in such a way that the zero columns of $\mathbf{C}(t)$ are grouped at the right-hand side of the matrix, and reorder the equations in such a way that the zero rows of $\mathbf{C}(t)$ are grouped in the lower part of the matrix, (5.38) becomes

$$\begin{bmatrix} \mathbf{A}_{11}(t) & \mathbf{A}_{12}(t) \\ \mathbf{A}_{21}(t) & \mathbf{A}_{22}(t) \end{bmatrix} \begin{bmatrix} \mathbf{X}_{n1} \\ \mathbf{X}_{n2} \end{bmatrix} dt + \begin{bmatrix} \mathbf{C}_{11}(t) & 0 \\ 0 & 0 \end{bmatrix} \begin{bmatrix} d\mathbf{X}_{n1} \\ d\mathbf{X}_{n2} \end{bmatrix}$$
$$+ \begin{bmatrix} \mathbf{B}_1(t) \\ \mathbf{B}_2(t) \end{bmatrix} d\mathbf{W}(t) = 0 \tag{5.41}$$

where $\mathbf{A}_{11}(t)$ and $\mathbf{C}_{11}(t)$ are $m \times m$, $\mathbf{A}_{22}(t)$ is $k \times k$, $\mathbf{A}_{12}(t)$ is $m \times k$, $\mathbf{A}_{21}(t)$ is $k \times m$, $\mathbf{B}_1(t)$ is $m \times p$, $\mathbf{B}_2(t)$ is $k \times p$, $\mathbf{X}_{n1}$ is an m-dimensional vector, $\mathbf{X}_{n2}$ is a k-dimensional vector, m is the number of nonzero columns (rows) in $C(t)$ and k is the number of zero columns (rows). Naturally, $n = m + k$. If we expand (5.41), we obtain two sets of equations, consisting of m and k equations:

$$\mathbf{A}_{11}(t)\,\mathbf{X}_{n1}\,dt + \mathbf{A}_{12}(t)\,\mathbf{X}_{n2}\,dt + \mathbf{C}_{11}(t)\,d\mathbf{X}_{n1} + \mathbf{B}_1(t)\,d\mathbf{W}(t) = 0, \tag{5.42}$$

$$\mathbf{A}_{21}(t)\,\mathbf{X}_{n1}\,dt + \mathbf{A}_{22}(t)\,\mathbf{X}_{n2}\,dt + \mathbf{B}_2(t)\,d\mathbf{W}(t) = 0. \tag{5.43}$$

We solve for $\mathbf{X}_{n2}\,dt$ in (5.43) to get

$$\mathbf{X}_{n2}\,dt = -[\mathbf{A}_{22}(t)^{-1}]\,\mathbf{A}_{21}(t)\,\mathbf{X}_{n1}\,dt - [\mathbf{A}_{22}(t)^{-1}]\,\mathbf{B}_2(t)\,d\mathbf{W}(t). \tag{5.44}$$

The above step assumes that $\mathbf{A}_{22}(t)$ ($k \times k$) is nonsingular. Nonsingularity of $\mathbf{A}_{22}(t)$ means that the variables in $\mathbf{X}_n$ which appear *without* time-derivatives in the equations can be expressed in terms of the variables which appear *with* derivatives. This condition is always satisfied if the variable is a node voltage. On the other hand, this condition is not always satisfied for voltage source currents, which always appear without derivatives in the equations. The only equations a voltage source current can appear in, are the KCL equations corresponding to the nodes the voltage source is connected to. If capacitors are connected to these nodes, then the only equations containing the voltage source current have derivatives in them, hence they can not be used to express the voltage source current in terms of the other variables, which means that $\mathbf{A}_{22}(t)$ is singular. Note that, $\mathbf{A}_{22}(t)$ becomes singular only if both nodes of the voltage source are connected to capacitors, or one of them is connected to a capacitor and the other is ground. This problem can be taken care of by eliminating the voltage source current variable and substituting the branch equation for this voltage source in the circuit equations in (5.38). At this point, we will assume that $\mathbf{A}_{22}(t)$ is nonsingular. Define

$$\begin{aligned}\mathbf{D}_1(t) &= -[\mathbf{A}_{22}(t)^{-1}]\,\mathbf{A}_{21}(t) \\ \mathbf{D}_2(t) &= -[\mathbf{A}_{22}(t)^{-1}]\,\mathbf{B}_2(t).\end{aligned} \tag{5.45}$$

Next, we use (5.44) and (5.45) in (5.42) to get

$$\begin{aligned}&\mathbf{A}_{11}(t)\,\mathbf{X}_{n1}\,dt + \mathbf{A}_{12}(t)\,\mathbf{D}_1(t)\,\mathbf{X}_{n1}\,dt \\ &+\mathbf{C}_{11}(t)\,d\mathbf{X}_{n1} + \mathbf{B}_1(t)\,d\mathbf{W}(t) + \mathbf{A}_{12}(t)\,\mathbf{D}_2(t)\,d\mathbf{W}(t) = 0\end{aligned} \tag{5.46}$$

and hence

$$\begin{aligned}\mathbf{C}_{11}(t)\,d\mathbf{X}_{n1} =\ &-[\mathbf{A}_{11}(t) + \mathbf{A}_{12}(t)\,\mathbf{D}_1(t)]\,\mathbf{X}_{n1}\,dt \\ &-[\mathbf{B}_1(t) + \mathbf{A}_{12}(t)\,\mathbf{D}_2(t)]\,d\mathbf{W}(t).\end{aligned} \tag{5.47}$$

Defining

$$\begin{aligned}\tilde{\mathbf{E}}(t) &= -(\mathbf{A}_{11}(t) + \mathbf{A}_{12}(t)\,\mathbf{D}_1(t)) \\ \tilde{\mathbf{F}}(t) &= -(\mathbf{B}_1(t) + \mathbf{A}_{12}(t)\,\mathbf{D}_2(t))\end{aligned} \tag{5.48}$$

and using (5.48) in (5.47) results in

$$\mathbf{C}_{11}(t)\,d\mathbf{X}_{n1} = \tilde{\mathbf{E}}(t)\,\mathbf{X}_{n1}\,dt + \tilde{\mathbf{F}}(t)\,d\mathbf{W}(t). \tag{5.49}$$

Now, we multiply both sides of (5.49) by the inverse of $\mathbf{C}_{11}(t)$ to obtain

$$d\mathbf{X}_{n1} = \mathbf{E}(t)\,\mathbf{X}_{n1}\,dt + \mathbf{F}(t)\,d\mathbf{W}(t) \tag{5.50}$$

where $\mathbf{E}(t) = [\mathbf{C}_{11}(t)]^{-1}\,\tilde{\mathbf{E}}(t)$ and $\mathbf{F}(t) = [\mathbf{C}_{11}(t)]^{-1}\,\tilde{\mathbf{F}}(t)$. $\mathbf{C}_{11}(t)$ is nonsingular provided that every node in the circuit with a connection to a capacitor has a *capacitive path* to ground or an independent voltage source node. We assume that this condition is satisfied by the circuit, hence $\mathbf{C}_{11}(t)$ is nonsingular. □

(5.50) is a system of linear stochastic differential equations in standard form. The right-hand-side in (5.50) is linear in $\mathbf{X}_{n1}$, and the coefficient matrices are independent of $\mathbf{X}_{n1}$. Just as with *linear* ordinary differential equations, a much more complete theory can be developed for *linear* stochastic differential equations [9]. As with ordinary differential equations, the general solution of a *linear* stochastic differential equation can be found explicitly. The method of solution also involves an integrating factor or, equivalently, a fundamental solution of an associated homogeneous differential equation.

Theorem 5.3 *The solution of (5.50) is given by*

$$\mathbf{X}_{n1}(t) = \mathbf{\Phi}(t,t_0)\,\mathbf{X}_{n1}(t_0) + \int_{t_0}^{t} \mathbf{\Phi}(t,\tau)\mathbf{F}(\tau)\,d\mathbf{W}(\tau) \tag{5.51}$$

where $\mathbf{\Phi}(t,\tau)$ is the state-transition matrix for the homogeneous differential equation

$$\dot{\mathbf{y}} = \mathbf{E}(t)\,\mathbf{y} \tag{5.52}$$

and hence can be determined as a function of t as the solution of

$$\frac{d\,\mathbf{\Phi}(t,\tau)}{dt} = \mathbf{E}(t)\,\mathbf{\Phi}(t,\tau)\,, \qquad \mathbf{\Phi}(\tau,\tau) = \mathbf{I}_m. \tag{5.53}$$

Proof. See [9]. Note that (5.51) involves a stochastic (Ito) integral. □

Corollary 5.1 *If the deterministic matrix functions $\mathbf{E}(t)$ and $\mathbf{F}(t)$ are bounded and piecewise-continuous in the time interval of interest, there exists a unique solution to (5.50) for every initial value vector $\mathbf{X}_{n1}(t_0)$.*

Proof. See [9]. □

The initial value vector $\mathbf{X}_{n1}(t_0)$ can either be a nonrandom constant vector or a vector of possibly correlated random variables.

Corollary 5.2 *If the initial value $\mathbf{X}_{n1}(t_0)$ is either a nonrandom constant vector or a vector of possibly correlated Gaussian random variables, then the solution $\mathbf{X}_{n1}(t)$ for $t \geq t_0$ of (5.50) is a vector of possibly nonstationary Gaussian processes.*

Proof. See [9]. This is a direct consequence of the fact that (5.50) represents a *linear* system.□

At this point, we state the following assumption:

Assumption 5.10 *The initial condition* $\mathbf{X}_{n1}(t_0)$ *is a vector of zero-mean Gaussian random variables.*

5.5 Derivation of Linear Time Varying ODEs for the Autocorrelation Matrix

We recall that the solution $\mathbf{X}(t)$ of (5.9) is approximated with

$$\mathbf{X}(t) \approx \mathbf{X}_0(t) + \epsilon \mathbf{X}_1(t) \tag{5.54}$$
$$\approx \mathbf{x}_s(t) + \epsilon \mathbf{X}_n(t). \tag{5.55}$$

With $\epsilon = 1$, (5.9) reduces to the original NLSDE (5.5) that describes a nonlinear circuit with the noise sources. Thus, we substitute $\epsilon = 1$ in (5.55) to obtain

$$\mathbf{X}(t) \approx \mathbf{x}_s(t) + \mathbf{X}_n(t). \tag{5.56}$$

We now would like to calculate the time evolution of the mean

$$\mathrm{E}\left[\mathbf{X}(t)\right] \tag{5.57}$$

and the autocorrelation function

$$\mathrm{E}\left[\mathbf{X}(t+\tau/2)\mathbf{X}(t-\tau/2)^T\right] \tag{5.58}$$

of $\mathbf{X}(t)$ for a complete second-order probabilistic characterization.

Theorem 5.4

$$\mathrm{E}\left[\mathbf{X}(t)\right] = \mathbf{x}_s(t) \tag{5.59}$$

where $\mathbf{x}_s(t)$ *is the solution of (5.3).*

Proof. By taking expectations of both sides of (5.51), we find out that

$$\mathrm{E}\left[\mathbf{X}_{n1}(t)\right] = \mathbf{0} \tag{5.60}$$

because $\mathrm{E}\left[\mathbf{X}_{n1}(t_0)\right] = \mathbf{0}$ as assumed, and $\mathrm{E}\left[\mathbf{W}(t)\right] = \mathbf{0}$ from the properties of the Wiener process. From (5.44) and (5.60), we conclude

$$\mathrm{E}\left[\mathbf{X}_{n2}(t)\right] = \mathbf{0}. \tag{5.61}$$

Then

$$\mathrm{E}\left[\mathbf{X}_n(t)\right] = \mathrm{E}\left[\begin{bmatrix} \mathbf{X}_{n1}(t) \\ \mathbf{X}_{n2}(t) \end{bmatrix}\right] = \mathbf{0} \tag{5.62}$$

follows from (5.60) and (5.61). Thus,

$$\mathrm{E}\left[\mathbf{X}(t)\right] = \mathbf{x}_s(t). \quad (5.63)$$

□

Using (5.62), the autocorrelation function of $\mathbf{X}(t)$ is given by

$$\mathrm{E}\left[\mathbf{X}(t+\tau/2)\mathbf{X}(t-\tau/2)^T\right] = \mathbf{x}_s(t+\tau/2)\,\mathbf{x}_s(t-\tau/2) \quad (5.64)$$
$$+ \mathrm{E}\left[\mathbf{X}_n(t+\tau/2)\mathbf{X}_n(t-\tau/2)^T\right]$$

in terms of $\mathbf{x}_s(t)$ and the autocorrelation function of $\mathbf{X}_n(t)$. So, we would like to calculate $\mathrm{E}\left[\mathbf{X}_n(t+\tau/2)\mathbf{X}_n(t-\tau/2)^T\right]$, which is given by

$$\mathrm{E}\left[\mathbf{X}_n(t+\tau/2)\mathbf{X}_n(t-\tau/2)^T\right] = \mathrm{E}\left[\begin{bmatrix} \mathbf{X}_{n1}(t+\tau/2) \\ \mathbf{X}_{n2}(t+\tau/2) \end{bmatrix} \begin{bmatrix} \mathbf{X}_{n1}(t-\tau/2) \\ \mathbf{X}_{n2}(t-\tau/2) \end{bmatrix}^T\right]$$

$$= \begin{bmatrix} \mathrm{E}\left[\mathbf{X}_{n1}(t+\tau/2)\mathbf{X}_{n1}(t-\tau/2)^T\right] & \mathrm{E}\left[\mathbf{X}_{n1}(t+\tau/2)\mathbf{X}_{n2}(t-\tau/2)^T\right] \\ \mathrm{E}\left[\mathbf{X}_{n2}(t+\tau/2)\mathbf{X}_{n1}(t-\tau/2)^T\right] & \mathrm{E}\left[\mathbf{X}_{n2}(t+\tau/2)\mathbf{X}_{n2}(t-\tau/2)^T\right] \end{bmatrix}.$$
(5.65)

Recall that the vector $\mathbf{X}_{n2}$ represents the noise component of the node voltages for the nodes which do *not* have a capacitive path to ground. For instance, the node between two resistors with no other connections is such a node. Since we are using idealized "white" noise models for thermal and shot noise, the variance of the node voltage for a node without a capacitive path the ground will be ∞! One way of intuitively seeing this is considering that the transfer functions from noise source currents in the circuit to a node without a capacitive path to ground will be "high-pass" functions of frequency, and hence the spectral density of this node voltage will not decay to zero as $f \to \infty$, resulting in infinite noise power. We rewrite (5.44) in the following form

$$\mathbf{X}_{n2}(t) = -[\mathbf{A}_{22}(t)^{-1}]\,\mathbf{A}_{21}(t)\,\mathbf{X}_{n1}(t) - [\mathbf{A}_{22}(t)^{-1}]\,\mathbf{B}_2(t)\,\frac{d\mathbf{W}(t)}{dt}. \quad (5.66)$$

Thus, $\mathbf{X}_{n2}(t)$ is a summation of two terms, and the second term is a modulated vector of white Gaussian processes (i.e. the formal derivative of the Wiener process). Hence, in general, all of the components of $\mathbf{X}_{n2}(t)$ have infinite variance, i.e. infinite power. So, it does not make sense to calculate $\mathrm{E}\left[\mathbf{X}_{n1}(t+\tau/2)\mathbf{X}_{n2}(t-\tau/2)^T\right]$, $\mathrm{E}\left[\mathbf{X}_{n2}(t+\tau/2)\mathbf{X}_{n1}(t-\tau/2)^T\right]$, or $\mathrm{E}\left[\mathbf{X}_{n2}(t+\tau/2)\mathbf{X}_{n2}(t-\tau/2)^T\right]$ in (5.65). We can only calculate the first diagonal block in (5.65), i.e. $\mathrm{E}\left[\mathbf{X}_{n1}(t+\tau/2)\mathbf{X}_{n1}(t-\tau/2)^T\right]$, the autocorrelation matrix of $\mathbf{X}_{n1}$. Recall that $\mathbf{X}_{n1}$ represents the noise component of the node voltages for the nodes which have a capacitive path to ground and the noise component of the inductor currents.

Theorem 5.5 *The time-varying variance-covariance matrix of* $\mathbf{X}_{n1}(t)$ *defined as*
$$\mathbf{K}(t) = \mathrm{E}\left[\mathbf{X}_{n1}(t)\mathbf{X}_{n1}(t)^T\right] \tag{5.67}$$
satisfies the following system of ordinary differential equations:
$$\frac{d\,\mathbf{K}(t)}{dt} = \mathbf{E}(t)\,\mathbf{K}(t) + \mathbf{K}(t)\,\mathbf{E}(t)^T + \mathbf{F}(t)\,\mathbf{F}(t)^T. \tag{5.68}$$

Proof. We start with calculating the following differential using Ito's formula (see Section 2.5.4)
$$d(\mathbf{X}_{n1}\mathbf{X}_{n1}^T) = (\mathbf{X}_{n1} + d\mathbf{X}_{n1})(\mathbf{X}_{n1}^T + d\mathbf{X}_{n1}^T) - \mathbf{X}_{n1}\mathbf{X}_{n1}^T. \tag{5.69}$$

From (5.69), it follows that
$$\begin{aligned} d(\mathbf{X}_{n1}\mathbf{X}_{n1}^T) &= \mathbf{X}_{n1}\,d\mathbf{X}_{n1}^T + (d\mathbf{X}_{n1})\,\mathbf{X}_{n1}^T + d\mathbf{X}_{n1}\,d\mathbf{X}_{n1}^T \\ &= \mathbf{X}_{n1}[\mathbf{X}_{n1}^T\,\mathbf{E}(t)^T\,dt + (d\mathbf{W}(t))^T\,\mathbf{F}(t)^T] \\ &\quad + [\mathbf{E}(t)\,\mathbf{X}_{n1}\,dt + \mathbf{F}(t)\,d\mathbf{W}(t)]\mathbf{X}_{n1}^T \\ &\quad + [\mathbf{E}(t)\,\mathbf{X}_{n1}\,dt + \mathbf{F}(t)\,d\mathbf{W}(t)] \\ &\quad [\mathbf{X}_{n1}{}^T\,\mathbf{E}(t)^T\,dt + (d\mathbf{W}(t))^T\,\mathbf{F}(t)^T] \end{aligned} \tag{5.70}$$

where we used (5.50). We expand (5.70) and neglect higher order terms according to (2.266), and obtain
$$\begin{aligned} d(\mathbf{X}_{n1}\mathbf{X}_{n1}^T) = &\,\mathbf{X}_{n1}\mathbf{X}_{n1}^T\,\mathbf{E}(t)^T\,dt + \mathbf{X}_{n1}(d\mathbf{W}(t))^T\,\mathbf{F}(t)^T + \\ &\,\mathbf{E}(t)\,\mathbf{X}_{n1}\mathbf{X}_{n1}^T\,dt + \mathbf{F}(t)\,(d\mathbf{W}(t))\mathbf{X}_{n1}^T + \\ &\,\mathbf{F}(t)\,d\mathbf{W}(t)\,(d\mathbf{W}(t))^T\,\mathbf{F}(t)^T. \end{aligned} \tag{5.71}$$

Then, we use (2.265) to substitute $(d\mathbf{W}(t))d\mathbf{W}(t)^T = \mathbf{I}_m dt$, and take the expectations of both sides of (5.71) to obtain
$$\begin{aligned} \mathrm{E}\left[d(\mathbf{X}_{n1}\mathbf{X}_{n1}^T)\right] = &\,\mathrm{E}\left[\mathbf{X}_{n1}\mathbf{X}_{n1}^T\right]\mathbf{E}(t)^T\,dt + \\ &\,\mathbf{E}(t)\,\mathrm{E}\left[\mathbf{X}_{n1}\mathbf{X}_{n1}^T\right]\,dt + \mathbf{F}(t)\,\mathbf{F}(t)^T\,dt. \end{aligned} \tag{5.72}$$

where we used the fact that $\mathbf{X}_{n1}(t)$ and $d\mathbf{W}(t)$ are uncorrelated, i.e.
$$\mathrm{E}\left[(d\mathbf{W}(t))\mathbf{X}_{n1}(t)^T\right] = 0, \tag{5.73}$$
$$\mathrm{E}\left[\mathbf{X}_{n1}(t)(d\mathbf{W}(t))^T\right] = 0. \tag{5.74}$$

This is merely a statement of the fact that the statistical properties of the white noise sources in the system at time t are independent of the past behavior of the system up to time t. Next, we substitute (5.67) in (5.72) and rewrite it in differential equation form:

$$\frac{d\,\mathbf{K}(t)}{dt} = \mathbf{E}(t)\,\mathbf{K}(t) + \mathbf{K}(t)\,\mathbf{E}(t)^T + \mathbf{F}(t)\,\mathbf{F}(t)^T. \tag{5.75}$$

□

(5.75) is a system of ordinary differential equations for the $m \times m$ time-varying variance-covariance matrix of $\mathbf{X}_{n1}(t)$ defined by (5.67). Being a variance-covariance matrix, $\mathbf{K}(t)$ is a symmetric and positive semi-definite matrix for all t.

Corollary 5.3 *If (5.75) satisfies the Lipschitz and boundedness conditions in the time interval of interest, a unique symmetric and positive semi-definite solution exists.*

In view of symmetry of $\mathbf{K}(t)$, (5.75) represents a system of $m(m+1)/2$ linear ordinary differential equations with time-varying coefficients. (5.75) can be solved numerically to calculate $\mathbf{K}(t)$ in the time interval of interest, given appropriate initial conditions. Choice of initial conditions and the numerical integration of (5.75) will be discussed in Section 5.7.

$\mathbf{K}(t)$ is the autocorrelation matrix of $\mathbf{X}_{n1}$, i.e. $\mathrm{E}\left[\mathbf{X}_{n1}(t+\tau/2)\mathbf{X}_{n1}(t-\tau/2)^T\right]$ evaluated at $\tau = 0$. For a complete second-order probabilistic characterization of $\mathbf{X}_{n1}$, we would like to be able to calculate $\mathrm{E}\left[\mathbf{X}_{n1}(t+\tau/2)\mathbf{X}_{n1}(t-\tau/2)^T\right]$ for $\tau \neq 0$ as well.

Theorem 5.6

$$\mathbf{K}(t_{ref}, t) = \mathrm{E}\left[\mathbf{X}_{n1}(t_{ref})\mathbf{X}_{n1}(t)^T\right] \tag{5.76}$$

satisfies the following system of ordinary differential equations

$$\frac{\partial\,\mathbf{K}(t_{ref}, t)}{\partial t} = \mathbf{K}(t_{ref}, t)\,\mathbf{E}(t)^T. \tag{5.77}$$

where t_{ref} is a fixed reference time point such that $t \geq t_{ref}$.

Proof. We start with calculating the following differential

$$d(\mathbf{X}_{n1}(t_{ref})\mathbf{X}_{n1}(t)^T) = \mathbf{X}_{n1}(t_{ref})(d\mathbf{X}_{n1}(t)^T). \tag{5.78}$$

From (5.78), it follows that

$$\begin{aligned} d(\mathbf{X}_{n1}(t_{ref})\mathbf{X}_{n1}(t)^T) &= \mathbf{X}_{n1}(t_{ref})[\mathbf{X}_{n1}(t)^T\,\mathbf{E}(t)^T\,dt + (d\mathbf{W}(t))^T\,\mathbf{F}(t)^T] \\ &= \mathbf{X}_{n1}(t_{ref})\,\mathbf{X}_{n1}(t)^T\,\mathbf{E}(t)^T\,dt \\ &\quad + \mathbf{X}_{n1}(t_{ref})\,(d\mathbf{W}(t))^T\,\mathbf{F}(t)^T. \end{aligned} \tag{5.79}$$

where we used (5.50). Then, we take the expectations of both sides of (5.79) to obtain

$$\mathrm{E}\left[d(\mathbf{X}_{n1}(t_{ref})\mathbf{X}_{n1}(t)^T)\right] = \mathrm{E}\left[\mathbf{X}_{n1}(t_{ref})\,\mathbf{X}_{n1}(t)^T\right]\mathbf{E}(t)^T\,dt \quad (5.80)$$

where we used the fact that $\mathbf{X}_{n1}(t_{ref})$ and $d\mathbf{W}(t)$ are uncorrelated. Next, we substitute (5.76) in (5.80) and rewrite it in differential equation form:

$$\frac{\partial\,\mathbf{K}(t_{ref},t)}{\partial t} = \mathbf{K}(t_{ref},t)\,\mathbf{E}(t)^T. \quad (5.81)$$

□

(5.81) is a system of ordinary differential equations for $\mathbf{K}(t_{ref},t)$ defined by (5.76). Integrating (5.81) for $t \geq t_{ref}$ at various values of t_{ref}, one can obtain a number of sections of the correlation matrix $\mathbf{K}(t_{ref},t)$. For these calculations, the initial condition for (5.81) is chosen as

$$\mathbf{K}(t_{ref},t)|_{t=t_{ref}} = \mathrm{E}\left[\mathbf{X}_{n1}(t_{ref})\mathbf{X}_{n1}(t_{ref})^T\right] \quad (5.82)$$

which is obtained from the solution of (5.75). Note that $\mathbf{K}(t_{ref},t)$ is not a symmetric matrix. The solution of (5.81) gives us $\mathbf{K}(t_1,t_2)$ for $t_2 \geq t_1$. To calculate $\mathbf{K}(t_1,t_2)$ for $t_2 < t_1$, one can use the simple formula

$$\mathbf{K}(t_1,t_2) = \mathbf{K}(t_2,t_1)^T. \quad (5.83)$$

By solving the system of ordinary differential equations given by (5.75) and (5.81), one can calculate $\mathrm{E}\left[\mathbf{X}_{n1}(t+\tau/2)\mathbf{X}_{n1}(t-\tau/2)^T\right]$, i.e. the autocorrelation function of $\mathbf{X}_{n1}$, for a range of t and τ that is of interest. Together with (5.60), we now have a method to calculate the complete second-order probabilistic characteristics of $\mathbf{X}_{n1}$ that represents the noise component of the node voltages for the nodes which have a capacitive path to ground, and the noise component of the inductor currents.

The derivations of (5.75) and (5.81) were based on (5.50), the system of linear stochastic differential equations in standard form we derived in Section 5.4. To obtain (5.50), we multiplied both sides of (5.49) by the inverse of $\mathbf{C}_{11}(t)$.[2] Recall that $\mathbf{C}_{11}(t)$ is the MNA matrix of the reactive elements, i.e. the capacitances and inductances in the circuit. For some circuits with widely varying capacitor values, even though $\mathbf{C}_{11}(t)$ is nonsingular, the numerical calculation of the coefficient matrices $\mathbf{E}(t)$ and $\mathbf{F}(t)$ in (5.50), which requires an LU decomposition of $\mathbf{C}_{11}(t)$, can become ill-conditioned. This was not an issue for the nonlinear circuits we analyzed. For broader applicability, we provide derivations of alternative systems of ordinary differential equations in Section 5.8. The equations that will be derived in Section 5.8 are based on the formulation (5.49) instead of (5.50), and hence these alternative derivations result in more numerically stable implementations.

5.6 Solution of the Linear Time Varying ODEs for the Autocorrelation Matrix

Theorem 5.7 *The solution of (5.75) is given by*

$$\mathbf{K}(t) = \mathbf{\Phi}(t,0)\mathbf{K}_0\mathbf{\Phi}(t,0)^T + \int_0^t \mathbf{\Phi}(t,\tau)\mathbf{F}(\tau)\mathbf{F}(\tau)^T \mathbf{\Phi}(t,\tau)^T d\tau \quad (5.84)$$

where $\mathbf{\Phi}(t,\tau)$ is the state transition matrix for the system of linear time-varying homogeneous differential equations

$$\dot{\mathbf{y}} = \mathbf{E}(t)\mathbf{y}. \quad (5.85)$$

$\mathbf{K}_0$ *in (5.84) denotes the initial condition* $\mathbf{K}(0) = \mathbf{K}_0$.

Proof. (5.84) is similar to (2.160), and it can be easily verified by substituting it into (5.75) [40].□

Theorem 5.8 *The solution of (5.81) is fiven by*

$$\mathbf{K}(t_{ref},t) = \mathbf{K}(t_{ref},t_{ref})\mathbf{\Phi}(t,t_{ref})^T \quad (5.86)$$

where the initial condition $\mathbf{K}(t_{ref}, t_{ref})$ *is obtained by solving (5.75).*

Obviously, we can not, in general, obtain analytical expressions for the state transition matrix $\mathbf{\Phi}(t, \tau)$ and hence for the solutions given in (5.84) and (5.86). We then revert to numerical methods to compute the solutions of (5.75) and (5.81), which we discuss in the next section along with some basic properties of the system of differential equations in (5.75) that is of interest to us.

5.7 Numerical Computation of the Autocorrelation Matrix

In this section, we describe how we numerically solve the system of ordinary differential equations in (5.75) and (5.81) to calculate $\mathbf{E}\left[\mathbf{X}_{n1}(t+\tau/2)\mathbf{X}_{n1}(t-\tau/2)^T\right]$, i.e. the autocorrelation function of $\mathbf{X}_{n1}$, for a range of t and τ that is of interest.

5.7.1 Computation of the coefficient matrices

To be able to solve (5.75), which we repeat below for convenience,

$$\frac{d\,\mathbf{K}(t)}{dt} = \mathbf{E}(t)\,\mathbf{K}(t) + \mathbf{K}(t)\,\mathbf{E}(t)^T + \mathbf{F}(t)\,\mathbf{F}(t)^T \quad (5.87)$$

we first need to calculate the coefficient matrices $\mathbf{E}(t)$ and $\mathbf{F}(t)$ in the time interval of interest $[0,T]$. The time-varying coefficient matrices $\mathbf{E}(t)$ and $\mathbf{F}(t)$

are not available analytically. To calculate $\mathbf{E}(t)$ and $\mathbf{F}(t)$ numerically, we first need to solve the nonlinear circuit equations without noise, i.e.

$$\mathbf{I}(\mathbf{x},t) + \frac{d}{dt}\mathbf{Q}(\mathbf{x}) = 0, \tag{5.88}$$

to calculate $\mathbf{x}_s(t)$ in the time interval of interest. $\mathbf{I}(\mathbf{x},t)$ and $\mathbf{Q}(\mathbf{x})$ in (5.88) are usually available as analytical expressions in the models of electronic components and semiconductor devices. The numerical calculation of $\mathbf{x}_s(t)$ is already implemented in the circuit simulator SPICE, and is called the *transient analysis*. The numerical methods for solving (5.88) subdivide the time interval of interest $[0,T]$ into a finite set of distinct points:

$$t_0 = 0, \quad t_R = T, \quad t_r = t_{r-1} + h_r \qquad r = 1, \ldots, R \tag{5.89}$$

where h_r are the time steps. At each time point t_r, the numerical methods compute an "approximation" $\mathbf{x}_s[r]$ to the exact solution $\mathbf{x}_s(t_r)$ [39].

Once we know $\mathbf{x}_s(t_r)$ at a specific time point t_r, we can now calculate $\mathbf{G}(t_r)$, $\mathbf{C}(t_r)$ and $\mathbf{B}(t_r)$ (all defined in Section 5.3) using (5.32), (5.19) and (5.26). The matrices $\mathbf{G}(t_r)$ and $\mathbf{C}(t_r)$ are basically the *Jacobians* of the vector functions $\mathbf{I}(\mathbf{x},t)$ and $\mathbf{Q}(\mathbf{x})$ evaluated at $\mathbf{x}_s(t_r)$, respectively. Since $\mathbf{I}(\mathbf{x},t)$ and $\mathbf{Q}(\mathbf{x})$ are available as analytical expressions, the Jacobians are also calculated analytically apriori (or with automatic differentiation methods), and they are included in the device models. $\mathbf{B}(t_r)$ is basically $\mathbf{B}(\mathbf{x}(t),t)$ (which contains the modulating functions for the noise sources that are available as analytical expressions as described in Chapter 3) evaluated at $\mathbf{x}_s(t_r)$. Then, $\mathbf{E}(t_r)$ and $\mathbf{F}(t_r)$ are calculated from $\mathbf{G}(t_r)$, $\mathbf{C}(t_r)$ and $\mathbf{B}(t_r)$ by performing the straightforward matrix operations described by the equations (5.36)-(5.50) in Section 5.4. All of these operations are performed with sparse matrix data structures. The numerical operations actually implemented somewhat differ from what has been described in Section 5.4 because of efficiency reasons.

The calculation of the matrix $\mathbf{A}(t) = \mathbf{G}(t) + \dot{\mathbf{C}}(t)$ in (5.38) requires the calculation of the time derivative $\dot{\mathbf{C}}(t)$. We compute $\dot{\mathbf{C}}(t)$ using the *same* time discretization formulas that were used by the transient analysis routines to discretize (5.88) in time to calculate $\mathbf{x}_s(t)$. For example, if Backward Euler is used to discretize (5.88), then

$$\dot{\mathbf{C}}(t_r) = \frac{\mathbf{C}(t_r) - \mathbf{C}(t_{r-1})}{h_r} \tag{5.90}$$

is used to calculate the time derivative of $\mathbf{C}(t_r)$ in calculating $\mathbf{A}(t_r)$ at the time point t_r. Alternatively, one can calculate the matrix $\dot{\mathbf{C}}(t)$ by "multiplying" the *Hessian* of $\mathbf{Q}(\mathbf{x})$ evaluated at $\mathbf{x}_s(t)$ with the vector $\dot{\mathbf{x}}_s(t)$, since $\mathbf{C}(t)$ is the

Jacobian of $\mathbf{Q}(\mathbf{x})$ evaluated at $\mathbf{x}_s(t)$. In this case, we again use the *same* time discretization formulas that were used by the transient analysis routines to discretize (5.88) to calculate $\dot{\mathbf{x}}_s(t)$. With this second method, one needs to calculate the second-order derivatives of $\mathbf{Q}(\mathbf{x})$, i.e. the Hessian, which has to be done apriori and the analytical expressions for the second-order derivatives have to be placed in the device models.

5.7.2 *Numerical solution of the differential Lyapunov matrix equation*

We now concentrate on the numerical integration of (5.87). In view of the symmetry of $\mathbf{K}(t)$ (which is a variance-covariance matrix), (5.87) represents a system of $m(m+1)/2$ linear ordinary differential equations with time-varying coefficients. The "matrix" differential equation (5.87) is in a special form. Matrix differential equations of this form appear in many engineering and mathematics problems in control theory, system theory, optimization, power systems, signal processing, etc. [40], and are referred to as the *differential Lyapunov matrix equation*. It is named after the Russian mathematician Alexander Mikhailovitch Lyapunov, who in 1892, introduced his famous stability theory for linear and nonlinear systems. In Lyapunov's stability theory for continuous-time systems, equations of the form (5.87) arise in the stability analysis of linear time-varying systems. The reader is referred to [40] for a discussion of problems where Lyapunov-like matrix equations appear.

Remark 5.5 *Given an initial condition, the time-varying matrix differential equation (5.87) has a unique solution under the condition that $\mathbf{E}(t)$ and $\mathbf{F}(t)$ are continuous and bounded time functions.*

Lemma 5.4 *The $m \times m$ matrix $\mathbf{F}(t)\,\mathbf{F}(t)^T$ in (5.87) is positive semidefinite for all t.*

Proof. See Section 2.4.4. □

Remark 5.6 *If the initial condition $\mathbf{K}(0) = \mathbf{K}_0$ is positive semidefinite, then the solution of (5.87), $\mathbf{K}(t)$, is positive semidefinite for $t \geq 0$.*

We recall that $\mathbf{K}(t)$ is the time-varying variance-covariance matrix for the vector of stochastic processes $\mathbf{X}_{n1}(t)$. As stated in Section 5.4:

Assumption 5.11 *We assume that $\mathbf{X}_{n1}(0)$ is a vector of zero-mean Gaussian random variables.*

Then,

Remark 5.7 $\mathbf{K}_0 = \mathrm{E}\left[\mathbf{X}_{n1}(0)\mathbf{X}_{n1}(0)^T\right]$ *is positive semidefinite, and hence $\mathbf{K}(t)$ is positive semidefinite for $t \geq 0$ as required since it is a variance-covariance matrix.*

For the numerical solution of (5.87), we discretize it in time using a suitable scheme. In general, one can use any linear multi-step method, or a Runge-Kutta method. For circuit simulation problems, implicit linear multi-step methods, and especially the trapezoidal method and the backward differentiation formula were found to be most suitable [39]. The trapezoidal method and the backward differentiation formula are almost exclusively used in circuit simulators in the numerical solution of (5.88). We use the trapezoidal method and the backward differentiation formula to discretize (5.87). Here, we will discuss the application of only backward Euler (backward differentiation formula of order 1). If backward Euler is applied to (5.87), we obtain

$$\frac{\mathbf{K}(t_r) - \mathbf{K}(t_{r-1})}{h_r} = \mathbf{E}(t_r)\,\mathbf{K}(t_r) + \mathbf{K}(t_r)\,\mathbf{E}(t_r)^T + \mathbf{F}(t_r)\,\mathbf{F}(t_r)^T. \quad (5.91)$$

One can put (5.91) in the below form by rearranging the terms

$$[\mathbf{E}(t_r) - \frac{\mathbf{I}_m}{2h_r}]\mathbf{K}(t_r) + \mathbf{K}(t_r)[\mathbf{E}(t_r) - \frac{\mathbf{I}_m}{2h_r}]^T + \mathbf{F}(t_r)\,\mathbf{F}(t_r)^T + \frac{\mathbf{K}(t_{r-1})}{h_r} = \mathbf{0}. \quad (5.92)$$

Let us define

$$\mathbf{P}_r = \mathbf{E}(t_r) - \frac{\mathbf{I}_m}{2h_r} \quad (5.93)$$

$$\mathbf{Q}_r = \mathbf{F}(t_r)\,\mathbf{F}(t_r)^T + \frac{\mathbf{K}(t_{r-1})}{h_r}. \quad (5.94)$$

With these definitions, (5.92) is rewritten as

$$\mathbf{P}_r\,\mathbf{K}(t_r) + \mathbf{K}(t_r)\,\mathbf{P}_r^T + \mathbf{Q}_r = \mathbf{0}. \quad (5.95)$$

$\mathbf{K}(t)$ at time point t_r is calculated by solving the system of linear equations in (5.95). In view of symmetry of $\mathbf{K}(t_r)$, (5.95) represents a system of $m(m+1)/2$ linear algebraic equations. The "matrix" equation (5.95) is in a special form, and algebraic matrix equations in this form are referred to as the *continuous-time algebraic Lyapunov matrix equation* in contrast with the differential Lyapunov matrix equation.

Lemma 5.5 *Let $\lambda_1, \lambda_2, \ldots, \lambda_m$ be the eigenvalues of the $m \times m$ matrix $\mathbf{P}_r$ in (5.95). (5.95) has a unique symmetric solution if and only if $\lambda_i + \lambda_j^* \neq 0$ for all $1 \leq i \leq m$ and $1 \leq j \leq m$.*

Proof. See [40]. □

The condition in Lemma 5.5 is obviously satisfied if every λ_i has a negative real part.

Lemma 5.6 *If every λ_i has a negative real part (i.e. $\mathbf{P}_r$ is "stable"), and if $\mathbf{Q}_r$ is positive semidefinite, then $\mathbf{K}(t_r)$ is also positive semidefinite.*

Proof. See [41]. □

Since $\mathbf{K}(t_r)$ is a variance-covariance matrix, it has to be positive semidefinite. During numerical integration, we have to make sure that this condition is satisfied at every time point, i.e. $\mathbf{K}(t_r)$ at every time point is a *valid* variance-covariance matrix.

Theorem 5.9 *If $\mathbf{K}(t_{r-1})$ calculated for the time point t_{r-1} is a valid variance-covariance matrix, i.e. it is positive semidefinite, then the time step h_r in (5.91) can be chosen in such a way to guarantee that $\mathbf{K}(t_r)$ for the time point t_r calculated as the solution of (5.95) is a valid variance-covariance matrix, i.e. it is positive semidefinite.*

Proof. Since $\mathbf{F}(t_r)\,\mathbf{F}(t_r)^T$ is always positive semidefinite, then $\mathbf{Q}_r$ is also positive semidefinite, because it is the summation of two positive semidefinite matrices. $\mathbf{E}(t_r)$ in (5.92) might have eigenvalues which have nonnegative real parts. This might be the case if the nonlinear circuit is an autonomous one, i.e. an oscillator. In nonautonomous circuits, there might be positive feedback loops in the circuit that become active for some period of time, e.g. during regenerative switchings. In order to guarantee that $\mathbf{K}(t_r)$ to be calculated is a valid variance-covariance matrix, all the eigenvalues of $\mathbf{P}_r$ should have negative real parts. This can be secured by choosing a small enough time step, because the eigenvalues of $\mathbf{P}_r$ are given by

$$\lambda_i = \mu_i - \frac{1}{2h_r} \quad 1 \leq i \leq m \tag{5.96}$$

in terms of the eigenvalues μ_i of $\mathbf{E}(t_r)$. Choosing "smaller" time steps during time intervals where $\mathbf{E}(t)$ have eigenvalues with nonnegative real parts also makes sense from an *accuracy* point of view, because $\mathbf{E}(t)$ having eigenvalues with nonnegative real parts suggests that the nonlinear circuit is in a fast "switching" positive feedback mode, which in turn means that the noise variances for the node voltages are also changing fast.□

The numerical integration of (5.87) can be performed concurrently with the numerical solution of (5.88). Since the solution of (5.88) is needed to calculate the coefficient matrices for (5.87), transient analysis is "forced" to solve for the time points that are needed during the numerical integration of (5.87). Ideally, the numerical integration of (5.88), i.e. transient analysis, and the numerical integration of (5.87), i.e. noise analysis, have separate automatic time step control mechanisms. There is, of course, an obvious correlation between the choice of time steps for (5.88) and (5.87), because the need to take smaller time steps for (5.88) suggests that smaller time steps are also necessary for (5.87).

We are solving (5.87) as an initial value problem, hence we need an initial condition $\mathbf{K}_0 = \mathbf{K}(0)$. We have to "choose" a positive semidefinite $\mathbf{K}_0$ as the initial condition. We set $\mathbf{K}_0$ to the solution of the following Lyapunov matrix equation

$$\mathbf{E}(0)\,\mathbf{K}_0 + \mathbf{K}_0\,\mathbf{E}(0)^T + \mathbf{F}(0)\,\mathbf{F}(0)^T = \mathbf{0} \tag{5.97}$$

whenever the solution is positive semidefinite. This will be the case if all of the eigenvalues of $\mathbf{E}(0)$ have negative real parts. If the solution of (5.97) is not positive semidefinite, we choose the initial condition as

$$\mathbf{K}_0 = \mathbf{0} \tag{5.98}$$

which is, for instance, the case for oscillator circuits. In Section 5.9, we will discuss the motivation behind choosing the initial condition as the solution of (5.97).

If any *implicit* linear multi-step method is applied to (5.87), we obtain a linear system of equations exactly in the form of (5.95). ($\mathbf{P}_r$ and $\mathbf{Q}_r$ will be given by different expressions for different implicit linear multi-step methods, but the resulting equation will always be in the form of (5.95).) We need to solve (5.95) at every time point. We discuss the numerical solution of (5.95) in the next section.

Above, we have only considered the numerical solution of (5.87). To be able to calculate the complete autocorrelation matrix of $\mathbf{X}_{n1}$, we also need to solve (5.81), which we repeat below

$$\frac{\partial\,\mathbf{K}(t_{ref},t)}{\partial t} = \mathbf{K}(t_{ref},t)\,\mathbf{E}(t)^T. \tag{5.99}$$

The initial condition for (5.99), $\mathbf{K}(t_{ref},t_{ref})$, is obtained by solving (5.87) from $t = 0$ to $t = t_{ref}$. If backward Euler is applied to (5.99), we obtain

$$\frac{\mathbf{K}(t_{ref},t_r) - \mathbf{K}(t_{ref},t_{r-1})}{h_r} = \mathbf{K}(t_{ref},t_r)\,\mathbf{E}(t_r)^T. \tag{5.100}$$

One can put (5.100) in the below form by rearranging the terms

$$[\mathbf{E}(t_r) - \frac{\mathbf{I}_m}{h_r}]\,\mathbf{K}(t_{ref},t_r)^T = -\frac{\mathbf{K}(t_{ref},t_{r-1})^T}{h_r}. \tag{5.101}$$

Then, one can solve (5.101) at a time point t_r for the columns of $\mathbf{K}(t_{ref},t_r)^T$ with one LU decomposition of $\mathbf{E}(t_r) - \frac{\mathbf{I}_m}{h_r}$ and m forward elimination and backward substitution steps.

In Section 5.8, we derive more numerically stable alternatives to (5.87) and (5.99) for circuits which have widely varying capacitor values that might result in numerical ill-conditioning when calculating the coefficient matrices $\mathbf{E}(t)$ and $\mathbf{F}(t)$. In Section 5.8, we also discuss the numerical solution of these more numerically stable alternatives to (5.87) and (5.99).

5.7.3 Numerical solution of the algebraic Lyapunov matrix equation

The first approach that comes to mind to solve (5.95) is to rewrite it as a sparse linear matrix-vector system in standard form and then use sparse matrix techniques for solving such systems. With this method, the system of equations (5.95) is converted into

$$\mathcal{A}\,\mathbf{y} = -\mathbf{b} \tag{5.102}$$

where

$$\begin{aligned}
\mathcal{A} &= \mathbf{I}_m \otimes \mathbf{P}_r + \mathbf{P}_r \otimes \mathbf{I}_m \\
\mathbf{y} &= [K_{11}, K_{21}, \ldots, K_{m1}, K_{12}, \ldots, K_{mm}]^T \\
\mathbf{b} &= [Q_{11}, Q_{21}, \ldots, Q_{m1}, Q_{12}, \ldots, Q_{mm}]^T.
\end{aligned} \tag{5.103}$$

K_{ij} and Q_{ij} in (5.103) denote the ijth entry of $\mathbf{K}(t_r)$ and $\mathbf{Q}_r$ respectively, and $\otimes$ denotes the Kronecker matrix product defined as

$$\mathbf{A} \otimes \mathbf{B} = \begin{bmatrix} a_{11}\mathbf{B} & a_{12}\mathbf{B} & \ldots & a_{1n}\mathbf{B} \\ a_{21}\mathbf{B} & a_{22}\mathbf{B} & \ldots & a_{2n}\mathbf{B} \\ \vdots & \vdots & & \vdots \\ a_{m1}\mathbf{B} & a_{m2}\mathbf{B} & \ldots & a_{mn}\mathbf{B} \end{bmatrix} \tag{5.104}$$

for an $m \times n$ matrix $\mathbf{A}$ and a matrix $\mathbf{B}$. $\mathcal{A}$ in (5.102) is a "very" sparse matrix. Even if the matrix $\mathbf{P}_r$ is full, $\mathcal{A}$ contains many entries which are structurally zero due to the special form of (5.95). Moreover, being similar to an MNA circuit matrix, $\mathbf{P}_r$ is also sparse which contributes to the sparsity of $\mathcal{A}$.

We used both a general-purpose direct method (i.e. sparse Gaussian elimination) sparse matrix solver, and an iterative sparse linear solver [42] (based on conjugate gradients squared) to solve the linear equation system in (5.102). The iterative solver performed significantly better than the direct solver, especially for equations obtained from larger circuits. Experiments with several circuits have shown that CPU time can be further reduced by using a parallel iterative linear solver (running on a CM-5) [43]. Parallel speed-ups with up to 50% efficiency were obtained with this parallel solver.

In the control theory literature, apart from the brute force method, several methods were proposed for the numerical solution of the algebraic Lyapunov matrix equation [40]. The Bartels-Stewart algorithm [44] was shown to be accurate and reliable. In this method, first $\mathbf{P}_r$ in (5.95) is reduced to upper Hessenberg form by means of Householder transformations, and then the QR-*algorithm* (not to be confused with the QR-*factorization*) is applied to the Hessenberg form to calculate the real Schur decomposition

$$\mathbf{S} = \mathbf{U}^T \mathbf{P}_r \mathbf{U} \tag{5.105}$$

of the matrix $\mathbf{P}_r$, where the real Schur form $\mathbf{S}$ is upper quasi-triangular (block upper triangular with 1×1 and 2×2 blocks on the diagonal, the eigenvalues of a 2×2 block being a complex conjugate pair), and $\mathbf{U}$ is orthonormal. The transformation matrices are accumulated at each step to form $\mathbf{U}$ [44]. Since reduction to Schur form is accomplished with *orthogonal* transformations, this process is numerically stable in contrast with, for instance, the computation of the Jordan form. Computation of the real Schur form of the matrix $\mathbf{P}_r$ requires $O(m^3)$ flops [44]. If we now set

$$\tilde{\mathbf{K}} = \mathbf{U}^T \mathbf{K}(t_r) \mathbf{U} \tag{5.106}$$
$$\tilde{\mathbf{Q}} = \mathbf{U}^T \mathbf{Q}_r \mathbf{U} \tag{5.107}$$

then (5.95) becomes

$$\mathbf{S}\tilde{\mathbf{K}} + \tilde{\mathbf{K}}\mathbf{S}^T = -\tilde{\mathbf{Q}}. \tag{5.108}$$

Solution of (5.108) for $\tilde{\mathbf{K}}$ can readily be accomplished by a process of forward substitution [44], because $\mathbf{S}$ is an upper quasi-triangular matrix. Once (5.108) is solved for $\tilde{\mathbf{K}}$, then $\mathbf{K}(t_r)$ can be computed using

$$\mathbf{K}(t_r) = \mathbf{U}\tilde{\mathbf{K}}\mathbf{U}^T \tag{5.109}$$

since $\mathbf{U}$ is orthonormal. The substitution phase also requires $O(m^3)$ flops [44]. In our experience, the Bartels-Stewart algorithm was computationally more efficient (for our specific problem) than the brute force methods described above. Moreover it requires less storage. It has very good numerical stability properties since it uses orthogonal transformations [45]. A variant of the Bartels-Stewart algorithm was proposed in [41] that allows the Cholesky factor of $\mathbf{K}(t_r)$ in (5.95) to be computed, without first computing $\mathbf{K}(t_r)$, when all of the eigenvalues of $\mathbf{P}_r$ have negative real parts and $\mathbf{Q}_r$ is positive semidefinite. Unfortunately, the Bartels-Stewart algorithm can not exploit the sparsity of the matrix $\mathbf{P}_r$ [46].

The numerical integration of (5.87) (with the Bartels-Stewart algorithm used to solve the algebraic Lyapunov matrix equation (5.95)) requires $O(m^3)$ flops at every time point compared with the roughly $O(m^{1.5})$ flops required by the numerical solution of (5.88). Hence, the CPU time usage will be largely dominated by the solution of (5.87). The computational cost would be high for "large" circuits, but this noise analysis method is intended for evaluating the noise performances of small (i.e. with several hundred state variables) subblocks (e.g. analog blocks such as mixers and oscillators) of a mixed-signal system design. Several iterative techniques have been proposed for the solution of the algebraic Lyapunov matrix equation (5.95) arising in some specific problems where the matrix $\mathbf{P}_r$ is large and sparse [46, 40]. The Krylov subspace based methods proposed in [47] and [48] seem to be promising. A matrix-implicit (without explicitly forming the matrix $\mathbf{P}_r$ in (5.95)) Krylov subspace based method with a specific preconditioner tuned for our problem to solve (5.95) seems to be a promising avenue to explore in an attempt to reduce the computational cost of the numerical solution of (5.87).

5.8 Alternative ODEs for the Autocorrelation Matrix[§]

We will use the formulation (5.49) for the derivations (Please see the discussion at the end of Section 5.5 for the motivation to derive these alternative ODEs for the autocorrelation matrix.) in this section, which is reproduced here for convenience:

$$\mathbf{C}_{11}(t)\, d\mathbf{X}_{n1} = \tilde{\mathbf{E}}(t)\, \mathbf{X}_{n1}\, dt + \tilde{\mathbf{F}}(t)\, d\mathbf{W}(t). \quad (5.110)$$

For notational simplicity, we now rewrite (5.111) with a change of symbols:

$$\mathbf{C}(t)\, d\mathbf{X}_{n1} = \mathbf{E}(t)\, \mathbf{X}_{n1}\, dt + \mathbf{F}(t)\, d\mathbf{W}(t). \quad (5.111)$$

5.8.1 Alternative ODE for the variance-covariance matrix

Theorem 5.10 *The time-varying variance-covariance matrix of $\mathbf{X}_{n1}(t)$ defined as*

$$\mathbf{K}(t) = \mathrm{E}\left[\mathbf{X}_{n1}(t)\mathbf{X}_{n1}(t)^T\right]. \quad (5.112)$$

satisfies the following system of ordinary differential equations

$$\mathbf{C}(t)\frac{d\,\mathbf{K}(t)}{dt}\mathbf{C}(t)^T = \mathbf{E}(t)\,\mathbf{K}(t)\,\mathbf{C}(t)^T + \mathbf{C}(t)\,\mathbf{K}(t)\,\mathbf{E}(t)^T + \mathbf{F}(t)\,\mathbf{F}(t)^T. \quad (5.113)$$

Proof. We start with calculating the following differential using Ito's formula (see Section 2.5.4)

$$\mathbf{C}(t)[d(\mathbf{X}_{n1}\mathbf{X}_{n1}^T)]\mathbf{C}(t)^T = \mathbf{C}(t)[(\mathbf{X}_{n1} + d\mathbf{X}_{n1})(\mathbf{X}_{n1}^T + d\mathbf{X}_{n1}^T) - \mathbf{X}_{n1}\mathbf{X}_{n1}^T]\mathbf{C}(t)^T. \quad (5.114)$$

From (5.114), it follows that

$$\begin{aligned}
\mathbf{C}(t)[d(\mathbf{X}_{n1}\mathbf{X}_{n1}^T)]\mathbf{C}(t)^T &= \mathbf{C}(t)[\mathbf{X}_{n1}\, d\mathbf{X}_{n1}^T + d\mathbf{X}_{n1}\, \mathbf{X}_{n1}^T \quad (5.115)\\
&\quad + d\mathbf{X}_{n1}\, d\mathbf{X}_{n1}^T]\mathbf{C}(t)^T\\
&= \mathbf{C}(t)\mathbf{X}_{n1} \quad (5.116)\\
&\quad [\mathbf{X}_{n1}^T\,\mathbf{E}(t)^T\, dt + (d\mathbf{W}(t))^T\,\mathbf{F}(t)^T]\\
&\quad + [\mathbf{E}(t)\,\mathbf{X}_{n1}\, dt + \mathbf{F}(t)\, d\mathbf{W}(t)]\mathbf{X}_{n1}^T\mathbf{C}(t)^T\\
&\quad + [\mathbf{E}(t)\,\mathbf{X}_{n1}\, dt + \mathbf{F}(t)\, d\mathbf{W}(t)]\\
&\quad [\mathbf{X}_{n1}^T\,\mathbf{E}(t)^T\, dt + (d\mathbf{W}(t))^T\,\mathbf{F}(t)^T]
\end{aligned}$$

[§]This section can be omitted without loss of continuity.

where we used (5.111). We expand (5.116) and neglect higher order terms according to (2.266), and obtain

$$\begin{aligned}\mathbf{C}(t)d(\mathbf{X}_{n1}\mathbf{X}_{n1}^T)\mathbf{C}(t)^T = & \mathbf{C}(t)\mathbf{X}_{n1}\mathbf{X}_{n1}^T\ \mathbf{E}(t)^T\ dt + \mathbf{C}(t)\mathbf{X}_{n1}(d\mathbf{W}(t))^T\ \mathbf{F}(t)^T + \\ & \mathbf{E}(t)\ \mathbf{X}_{n1}\mathbf{X}_{n1}^T\mathbf{C}(t)^T\ dt + \mathbf{F}(t)\ (d\mathbf{W}(t))\mathbf{X}_{n1}^T\mathbf{C}(t)^T + \\ & \mathbf{F}(t)\ d\mathbf{W}(t)\ (d\mathbf{W}(t))^T\ \mathbf{F}(t)^T. \end{aligned}$$
(5.117)

Then, we use (2.265) and substitute $(d\mathbf{W}(t))d\mathbf{W}(t)^T = \mathbf{I}_m dt$ and take the expectations of both sides of (5.117) to obtain

$$\begin{aligned}\mathbf{C}(t)\mathrm{E}\left[d(\mathbf{X}_{n1}\mathbf{X}_{n1}^T)\right]\mathbf{C}(t)^T = & \mathbf{C}(t)\mathrm{E}\left[\mathbf{X}_{n1}\mathbf{X}_{n1}^T\right]\ \mathbf{E}(t)^T\ dt \\ & + \mathbf{E}(t)\ \mathrm{E}\left[\mathbf{X}_{n1}\mathbf{X}_{n1}^T\right]\mathbf{C}(t)^T\ dt \\ & + \mathbf{F}(t)\ \mathbf{F}(t)^T\ dt \end{aligned} \quad (5.118)$$

where we used the fact that $\mathbf{X}_{n1}(t)$ and $d\mathbf{W}(t)$ are uncorrelated. Next, we substitute (5.112) in (5.118) and rewrite it in differential equation form:

$$\mathbf{C}(t)\frac{d\ \mathbf{K}(t)}{dt}\mathbf{C}(t)^T = \mathbf{E}(t)\ \mathbf{K}(t)\ \mathbf{C}(t)^T + \mathbf{C}(t)\ \mathbf{K}(t)\ \mathbf{E}(t)^T + \mathbf{F}(t)\ \mathbf{F}(t)^T. \quad (5.119)$$
□

(5.119), the alternative to (5.75), is a system of ordinary differential equations for the $m \times m$ time-varying variance-covariance matrix of $\mathbf{X}_{n1}(t)$ defined by (5.112).

5.8.2 Alternative ODE for the correlation matrix

Theorem 5.11

$$\mathbf{K}(t_{ref}, t) = \mathrm{E}\left[\mathbf{X}_{n1}(t_{ref})\mathbf{X}_{n1}(t)^T\right] \quad (5.120)$$

satisfies the following system of ordinary differential equations

$$\mathbf{C}(t_{ref})\frac{\partial\ \mathbf{K}(t_{ref}, t)}{\partial t}\mathbf{C}(t)^T = \mathbf{C}(t_{ref})\mathbf{K}(t_{ref}, t)\ \mathbf{E}(t)^T \quad (5.121)$$

where t_{ref} is a fixed reference time point such that $t \geq t_{ref}$.

Proof. We start with calculating the following differential

$$\mathbf{C}(t_{ref})[d(\mathbf{X}_{n1}(t_{ref})\mathbf{X}_{n1}(t)^T)]\mathbf{C}(t)^T = \mathbf{C}(t_{ref})[\mathbf{X}_{n1}(t_{ref})(d\mathbf{X}_{n1}(t)^T)]\mathbf{C}(t)^T. \quad (5.122)$$

From (5.122), it follows that

$$\begin{aligned}\mathbf{C}(t_{ref})d(\mathbf{X}_{n1}(t_{ref})\mathbf{X}_{n1}(t)^T)\mathbf{C}(t)^T = & \mathbf{C}(t_{ref})\mathbf{X}_{n1}(t_{ref}) \\ & [\mathbf{X}_{n1}(t)^T\mathbf{E}(t)^T\ dt \\ & + d\mathbf{W}(t)^T\mathbf{F}(t)^T]\end{aligned}$$

$$= \mathbf{C}(t_{ref})\mathbf{X}_{n1}(t_{ref}) \quad (5.123)$$
$$\mathbf{X}_{n1}(t)^T \mathbf{E}(t)^T dt$$
$$+\mathbf{C}(t_{ref})\mathbf{X}_{n1}(t_{ref}) (d\mathbf{W}(t))^T \mathbf{F}(t)^T$$

where we used (5.111). Then, we take the expectations of both sides of (5.123) to obtain

$$\mathbf{C}(t_{ref})\mathrm{E}\left[d(\mathbf{X}_{n1}(t_{ref})\mathbf{X}_{n1}(t)^T)\right]\mathbf{C}(t)^T = \\ \mathbf{C}(t_{ref})\mathrm{E}\left[\mathbf{X}_{n1}(t_{ref})\mathbf{X}_{n1}(t)^T\right]\mathbf{E}(t)^T dt \quad (5.124)$$

where we used the fact that $\mathbf{X}_{n1}(t_{ref})$ and $d\mathbf{W}(t)$ are uncorrelated. Next, we substitute (5.120) in (5.124) and rewrite it in differential equation form:

$$\mathbf{C}(t_{ref})\frac{\partial \mathbf{K}(t_{ref},t)}{\partial t}\mathbf{C}(t)^T = \mathbf{C}(t_{ref})\mathbf{K}(t_{ref},t) \mathbf{E}(t)^T \quad (5.125)$$

□

Since $\mathbf{C}(t_{ref})$ is a constant nonsingular matrix, (5.125) is equivalent to

$$\frac{\partial \mathbf{K}(t_{ref},t)}{\partial t}\mathbf{C}(t)^T = \mathbf{K}(t_{ref},t) \mathbf{E}(t)^T. \quad (5.126)$$

(5.126), the alternative to (5.81), is a system of ordinary differential equations for $\mathbf{K}(t_{ref},t)$ defined by (5.120).

5.8.3 Numerical computation of the autocorrelation matrix

If backward Euler is applied to (5.119), we obtain

$$\mathbf{C}(t_r)[\frac{\mathbf{K}(t_r) - \mathbf{K}(t_{r-1})}{h_r}]\mathbf{C}(t_r)^T = \mathbf{E}(t_r) \mathbf{K}(t_r) \mathbf{C}(t_r)^T + \mathbf{C}(t_r) \mathbf{K}(t_r) \mathbf{E}(t_r)^T \\ +\mathbf{F}(t_r) \mathbf{F}(t_r)^T \quad (5.127)$$

One can put (5.127) in the below form by rearranging the terms

$$[\mathbf{E}(t_r) - \frac{\mathbf{C}(t_r)}{2h_r}] \mathbf{K}(t_r) \mathbf{C}(t_r)^T + \mathbf{C}(t_r) \mathbf{K}(t_r) [\mathbf{E}(t_r) - \frac{\mathbf{C}(t_r)}{2h_r}]^T \\ +\mathbf{F}(t_r) \mathbf{F}(t_r)^T + \mathbf{C}(t_r) \frac{\mathbf{K}(t_{r-1})}{h_r} \mathbf{C}(t_r)^T = \mathbf{0}. \quad (5.128)$$

Let us define

$$\mathbf{P}_r = \mathbf{E}(t_r) - \frac{\mathbf{C}(t_r)}{2h_r} \quad (5.129)$$

$$\mathbf{Q}_r = \mathbf{F}(t_r) \mathbf{F}(t_r)^T + \mathbf{C}(t_r) \frac{\mathbf{K}(t_{r-1})}{h_r} \mathbf{C}(t_r)^T \quad (5.130)$$

$$\mathbf{C}_r = \mathbf{C}(t_r) \quad (5.131)$$

144   NOISE IN NONLINEAR ELECTRONIC CIRCUITS

With these definitions, (5.128) is rewritten as

$$\mathbf{P}_r \, \mathbf{K}(t_r) \, \mathbf{C}_r^T + \mathbf{C}_r \, \mathbf{K}(t_r) \, \mathbf{P}_r^T + \mathbf{Q}_r = \mathbf{0}. \tag{5.132}$$

$\mathbf{K}(t)$ at time point t_r is calculated by solving the system of linear equations in (5.132). In view of the symmetry of $\mathbf{K}(t_r)$, (5.132) represents a system of $m(m+1)/2$ linear algebraic equations. The "matrix" equation (5.132) is in a special form. When $\mathbf{C}_r$ is nonsingular (which is assumed to be true), (5.132) can be put into the form of the continuous-time algebraic Lyapunov matrix equation given in (5.95), but this would defeat the whole purpose of deriving the alternative ODE (5.119), for which the coefficient matrices can be computed in a numerically stable way. Thus, when $\mathbf{C}_r$ is poorly conditioned, we avoid the transformation of (5.132) into the form of (5.95) and solve (5.132) directly at every time point using an algorithm that is similar to the Bartels-Stewart algorithm [45]. . If any *implicit* linear multi-step method is applied to (5.119), we obtain a linear system of equations exactly in the form of (5.132).

If backward Euler is applied to (5.126), we obtain

$$[\frac{\mathbf{K}(t_{ref}, t_r) - \mathbf{K}(t_{ref}, t_{r-1})}{h_r}] \mathbf{C}(t_r) = \mathbf{K}(t_{ref}, t_r) \, \mathbf{E}(t_r)^T. \tag{5.133}$$

One can put (5.133) in the below form by rearranging the terms

$$[\mathbf{E}(t_r) - \frac{\mathbf{C}(t_r)^T}{h_r}] \, \mathbf{K}(t_{ref}, t_r)^T = -\mathbf{C}(t_r)^T \, \frac{\mathbf{K}(t_{ref}, t_{r-1})^T}{h_r}. \tag{5.134}$$

Then, one can solve (5.134) at time t_r for the columns of $\mathbf{K}(t_{ref}, t_r)^T$ with one LU decomposition of $[\mathbf{E}(t_r) - \frac{\mathbf{C}(t_r)^T}{h_r}]$ and m forward elimination and backward substitution steps.

Thus, with the formulation described in this section, we avoid calculating the LU decomposition of $\mathbf{C}(t)$ which might be a numerically ill-conditioned operation for circuits with widely varying capacitor values.

5.9 Time-Invariant and Periodic Steady-State

We now concentrate on nonlinear circuits with a large signal time-invariant or periodic steady-state solution, and hence the time-invariant or periodic steady-state solutions of (5.87) and (5.99).

5.9.1 Time-invariant steady-state

In general, for a nonlinear dynamic circuit with arbitrary time-varying excitations, $\mathbf{E}(t)$ and $\mathbf{F}(t)$ in (5.87) are arbitrary time-varying functions. On the other hand, for nonlinear circuits with a time-invariant large-signal steady-state solution (i.e. (5.88) has a steady-state solution $\mathbf{x}_s(t)$ that is a constant function of time), $\mathbf{E}(t)$ and $\mathbf{F}(t)$ are constant functions of time. We assume that the nonlinear circuit is in time-invariant large-signal steady-state at $t = 0$, that is,

$$\begin{aligned} \mathbf{E}(t) &= \mathbf{E}(0) \\ \mathbf{F}(t) &= \mathbf{F}(0) \end{aligned} \tag{5.135}$$

for $t \geq 0$. In this case, with the initial value $\mathbf{K}_0$ being set to the solution of (5.97) (when there is a positive semidefinite solution), the solution of (5.87) is given by

$$\mathbf{K}(t) = \mathbf{K}_0 \qquad t \geq 0. \tag{5.136}$$

Thus, the initial condition $\mathbf{K}_0$ calculated as the solution of (5.97) is basically the time-invariant steady-state solution (when there exists one) of (5.87). In steady-state, $\mathbf{X}_{n1}(t)$ has a *time-invariant* variance-covariance matrix. Thus,

Remark 5.8 *If there exists a positive semidefinite solution to (5.97), and if (5.135) is satisifed, then*

- *(5.87) has the time-invariant positive semidefinite steady-state solution given by (5.136) where $\mathbf{K}_0$ is the positive semidefinite solution of (5.97).*

- $\mathbf{X}_{n1}(t)$ *(the noise component of the node voltages for the nodes which have a capacitive path to ground and the noise component of the inductor currents) is a vector of (asymptotically) WSS stochastic processes.*

Note that the existence of a large-signal time-invariant steady-state solution for (5.88) does not guarantee that there exists a *positive semidefinite* solution for (5.97) and hence a positive semidefinite steady-state solution for (5.87).

Remark 5.9 *There exists a time-invariant postive semidefinite steady-state solution for (5.87) if all of the eigenvalues of $\mathbf{E}(t) = \mathbf{E}(0)$ have negative real parts, i.e. if the circuit is "stable".*

In this case, the autocorrelation matrix for the WSS $\mathbf{X}_{n1}(t)$, i.e.

$$\mathbf{R}_{\mathbf{X}_{n1}}(\tau) = \mathbf{E}\left[\mathbf{X}_{n1}(t+\tau)\mathbf{X}_{n1}(t)^T\right] \tag{5.137}$$

can be calculated by solving (5.99). The initial condition for the system of homogeneous differential equations (5.99), i.e. $\mathbf{R}_{\mathbf{X}_{n1}}(0)$, is the steady-state time-invariant solution of (5.87) given in (5.136). In general, for a "stable" circuit, the autocorrelation function $\mathbf{R}_{\mathbf{X}_{n1}}(\tau)$ satisfies

$$\mathbf{R}_{\mathbf{X}_{n1}}(\tau) \to 0 \quad \text{as} \quad \tau \to \pm \infty. \tag{5.138}$$

The time-domain noise simulation for a "stable" circuit in time-invariant large-signal steady-state reduces to solving the linear equation system (5.97), and then solving (5.99) to calculate the autocorrelation functions for the WSS output. This can be compared with the traditional frequency domain noise analysis based on LTI transformations (see Section 4.2) which works for circuits in time-invariant steady-state, and calculates the spectral densities. Solving (5.97) is equivalent to calculating the total integrated noise for all of the circuit variables over the frequency range from 0 to ∞ (that is, the noise variances for all of the circuit variables computed by calculating the integral of the spectral densities) in SPICE AC noise analysis. In fact, the solution of (5.97) also provides the noise covariances for all of the circuit variables. Calculating the noise covariance between two circuit variables in SPICE noise simulation requires the calculation of total integrated noise over the frequency range from 0 to ∞ for the difference of the two circuit variables.

5.9.2 Periodic steady-state

For nonlinear circuits with a periodic large-signal steady-state solution (i.e. (5.88) has a steady-state solution $\mathbf{x}_s(t)$ that is a periodic function of time), $\mathbf{E}(t)$ and $\mathbf{F}(t)$ in (5.87) are periodic functions of time. We assume that the nonlinear circuit is in periodic large-signal steady-state (if there exists one), that is,

$$\mathbf{E}(t + kT) = \mathbf{E}(t) \qquad (5.139)$$
$$\mathbf{F}(t + kT) = \mathbf{F}(t) \qquad (5.140)$$

for all t, and $k \in \mathbf{Z}$, and for some period $T > 0$. The existence of a large-signal periodic steady-state solution for (5.88) does *not* necessarily mean that there exists a *positive semidefinite* periodic steady-state solution for (5.87). In a loose sense, there exists a periodic steady-state solution for (5.87) if the nonlinear circuit is nonautonomous and the large-signal periodic steady-state is forced with some periodic excitations. To make this condition precise, we state

Remark 5.10 *There exists a periodic steady-state solution for (5.87), if the Floquet exponents[3] that are associated with the periodically time-varying system matrix $\mathbf{E}(t)$ have strictly negative real parts.*

When (5.87) has a periodic steady-state solution, the initial condition $\mathbf{K}_0$ for (5.87), which will enable us to calculate the periodic steady-state solution for (5.87) by numerically integrating it from $t = 0$ to $t = T$, satisfies the equation

$$\mathbf{K}_0 = \mathbf{\Phi}(T, 0) \mathbf{K}_0 \mathbf{\Phi}(T, 0)^T + \int_0^T \mathbf{\Phi}(t, \tau) \mathbf{F}(\tau) \mathbf{F}(\tau)^T \mathbf{\Phi}(t, \tau)^T d\tau \qquad (5.141)$$

which was obtained from (5.84) by setting $\mathbf{K}(T) = \mathbf{K}_0$. We can rewrite (5.141) as follows:
$$\mathbf{K}_0 - \mathbf{\Phi}(T,0)\mathbf{K}_0\mathbf{\Phi}(T,0)^T = \mathbf{K}_p(T) \tag{5.142}$$
where
$$\mathbf{K}_p(T) = \int_0^T \mathbf{\Phi}(t,\tau)\mathbf{F}(\tau)\mathbf{F}(\tau)^T\mathbf{\Phi}(t,\tau)^T d\tau \tag{5.143}$$
which can be calculated by numerically integrating (5.87) with an initial condition $\mathbf{K}(0) = \mathbf{0}$. $\mathbf{\Phi}(T,0)$ can be calculated by numerically integrating (5.85) with an initial condition $\mathbf{y}(0) = \mathbf{I}_m$. (5.142) is a system of algebraic linear equations for the entries of the matrix $\mathbf{K}_0$. Algebraic matrix equations of the form (5.142) are referred to as *discrete-time algebraic Lyapunov matrix equations*, because they arise in Lyapunov's stability theory for discrete-time systems.

Lemma 5.7 *(5.142) has a unique symmetric solution if and only if $\mu_i \mu_j^* \neq 1$ for all $1 \leq i \leq m$ and $1 \leq j \leq m$, where $\mu_1, \mu_2, \ldots, \mu_m$ are the eigenvalues of the $m \times m$ matrix $\mathbf{\Phi}(T,0)$.*

Proof. See [41]. □

Recall that the eigenvalues μ_i of $\mathbf{\Phi}(T,0)$ are related to the Floquet exponents η_i that are associated with the periodic $\mathbf{E}(t)$ as follows:
$$\mu_i = \exp(\eta_i T). \tag{5.144}$$

Thus, the condition in Lemma 5.7 on the eigenvalues is obviously satisfied if all of the Floquet exponents have negative real parts. Moreover,

Lemma 5.8 *If all of the Floquet exponents that are associated with the periodic $\mathbf{E}(t)$ have negative real parts, (5.142) has a positive semidefinite solution.*

Proof. This is shown by observing that $\mathbf{K}_p(T)$ is positive semidefinite [41]. □

(5.142) can be numerically solved with an algorithm that is similar to the Bartels-Stewart algorithm for the continuous-time algebraic Lyapunov matrix equation [45].

When there exists a periodic steady-state solution for (5.87), $\mathbf{X}_{n1}(t)$ has a *periodically time-varying* variance-covariance matrix. Thus, it can be shown that $\mathbf{X}_{n1}(t)$ (the noise component of the node voltages for the nodes which have a capacitive path to ground and the noise component of the inductor currents) is a vector of (asymptotically) *cyclostationary* stochastic processes. In this case, the solution of (5.99) for a nonautonomous nonlinear circuit in periodic steady-state (with Floquet exponents that all have negative real parts), i.e. $\mathbf{K}(t_{ref}, t)$, satisfies
$$\mathbf{K}(t_{ref}, t) \to 0 \quad \text{as} \quad (t - t_{ref}) \to \pm\infty. \tag{5.145}$$

148   NOISE IN NONLINEAR ELECTRONIC CIRCUITS

As we will see in Chapter 6, for autonomous circuits with a large-signal periodic steady-state solution for (5.88), e.g. oscillators, one of the Floquet exponents associated with the periodic system matrix $\mathbf{E}(t)$ is exactly equal to zero. Hence, (5.87) does *not* have a periodic steady-state solution. Moreover, the solution of (5.87) with an initial condition $\mathbf{K}(0) = \mathbf{0}$ becomes unbounded as $t \to \infty$. Thus, $\mathbf{X}_{n1}(t)$ is not a vector of cyclostationary stochastic processes. In this case, (5.99) has a periodic steady-state solution for nonzero initial conditions. In Chapter 6, we will deeply investigate the meaning of these results for autonomous nonlinear circuits with a large-signal periodic steady-solution for (5.88).

5.10   Examples

We implemented the numerical computation of the autocorrelation matrix, as described in Section 5.7, in the circuit simulator SPICE [28, 49]. We now present several examples[4] of noise simulation using this implementation of the *non-Monte Carlo time-domain* technique we presented in this chapter.

5.10.1   Parallel RLC circuit

We start with a simple parallel RLC circuit. The values for the components are

$$\begin{aligned} R &= 1\ K\Omega \\ C &= 1\ pF \\ L &= 1\ \mu H. \end{aligned} \qquad (5.146)$$

This is obviously a simple LTI circuit and the noise analysis can be accomplished using the technique described in Section 4.2, but it illustrates what one can compute with the time-domain noise simulation technique described in Chapter 5. The noise source in the circuit is the current noise source that models the thermal noise of the resistor. We computed the variance of the noise voltage across the parallel RLC circuit, i.e. $\mathbf{E}\left[V_n(t)^2\right]$, as a function of time, which can be seen in Figure 5.1.[5] For this simulation, the initial condition for the noise voltage variance was set as $\mathbf{E}\left[V_n(0)^2\right] = 0$. As seen in Figure 5.1, the variance settles to a time-invariant steady-state value. Thus, at steady-state, the noise voltage across the parallel RLC circuit is a *WSS* process. We can also compute the autocorrelation function $R_{V_n}(\tau) = \mathbf{E}\left[V_n(t)V_n(t+\tau)\right]$ for this WSS process, which can be seen in Figure 5.2.[6] Note that $R_{V_n}(0)$ in Figure 5.2 is the steady-state value of $\mathbf{E}\left[V_n(t)^2\right]$ in Figure 5.1. Since the parallel RLC circuit (with $R > 0$) is a stable system, the autocorrelation function $R_{V_n}(\tau)$ satisfies

$$R_{V_n}(\tau) \to 0 \quad \text{as} \quad \tau \to \pm \infty. \qquad (5.147)$$

TIME-DOMAIN NON-MONTE CARLO NOISE SIMULATION 149

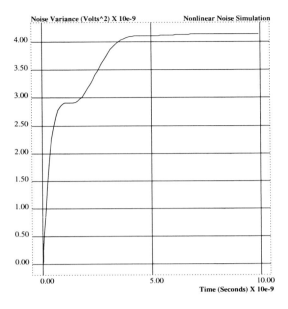

Figure 5.1. Noise voltage variance for the parallel RLC circuit

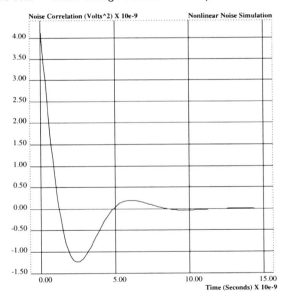

Figure 5.2. Noise voltage autocorrelation for the parallel RLC circuit

5.10.2 Switching noise of an inverter

For a CMOS inverter that is loaded with a 1 pF capacitor, and driven with a periodic large-signal waveform at the input, a noise simulation was performed. The large-signal waveform (obtained by transient analysis in SPICE) and the voltage noise variance waveform $\mathbf{E}\left[V_n(t)^2\right]$ (obtained from the solution of (5.87) with initial condition set to the solution of (5.97)) at the output of this inverter can be seen in Figure 5.3. We conclude from Figure 5.3 that the noise at the output is in general *nonstationary*[7], because the noise variance is not a constant as a function of time. We also observe that noise variance (i.e. mean-squared noise power) is highest during the low-to-high and high-to-low transitions of the large-signal output waveform. It is possible to calculate the *timing jitter* of the transitions at the output using the information in Figure 5.3. We will deeply explore the notion of *timing jitter* and its characterization in Chapter 6.

We can also compute the correlation function $R_{V_n}(t_{ref}, t) = \mathbf{E}\left[V_n(t_{ref})V_n(t)\right]$ for the noise voltage at the output by solving (5.99). Since the noise voltage is a nonstationary process, $R_{V_n}(t_{ref}, t)$ is also a function of a reference time point t_{ref}. For the simulation shown in Figure 5.4, the reference time point was chosen as $t_{ref} = 15$ $nsecs$. The initial condition $R_{V_n}(t_{ref}, t_{ref})$ is obtained from the simulation in Figure 5.3. The correlation function $R_{V_n}(t_{ref}, t)$ for the CMOS inverter satisfies

$$R_{V_n}(t_{ref}, t) \to 0 \quad \text{as} \quad (t - t_{ref}) \to \pm\infty \tag{5.148}$$

because it is a "stable" circuit.

5.10.3 Mixer noise figure

This bipolar Gilbert mixer circuit contains 14 BJTs, 21 resistors, 5 capacitors, and 18 parasitic capacitors connected between some of the nodes and ground. The LO (local oscillator) input is a sine-wave at 1.75 GHz with an amplitude of 178 mV. The RF input is a sine-wave at 2 GHz with an amplitude of 31.6 mV. Thus, the IF frequency is 250 MHz. With the above RF and LO inputs, the AC coupled IF output, obtained by transient analysis, is shown in Figure 5.5. This circuit was simulated to compute the noise voltage variance $\mathbf{E}\left[V_n(t)^2\right]$ at the IF output as a function of time, which can be seen in Figure 5.6. This noise variance waveform is periodic in steady-state with a period of 4 $nsecs$ (IF frequency is 250 MHz.), because the circuit is being driven with two periodic excitations that have commensurate frequencies. The noise voltage at the IF output of this circuit is *nonstationary*, because the signals applied to the circuit are large enough to change the operating point.

TIME-DOMAIN NON-MONTE CARLO NOISE SIMULATION 151

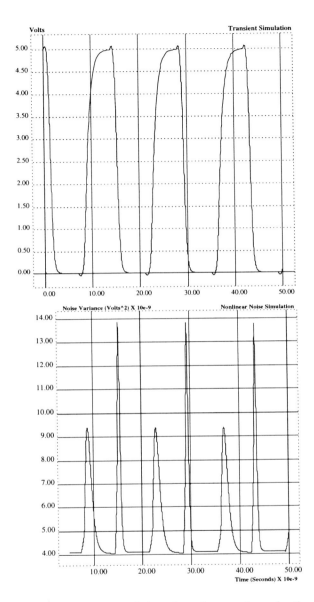

Figure 5.3. Large-signal waveform and the noise voltage variance for the CMOS inverter

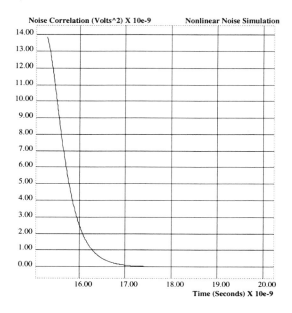

Figure 5.4. Noise voltage correlation for the CMOS inverter

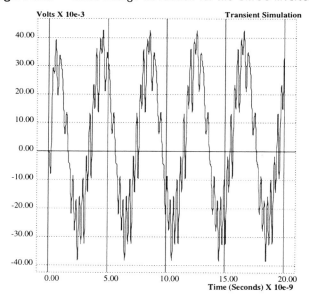

Figure 5.5. Large signal IF output for the mixer

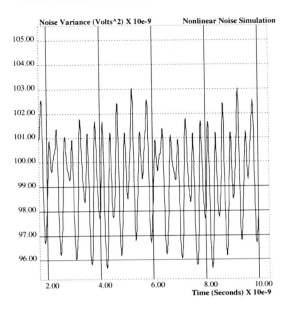

Figure 5.6. Noise voltage variance at the IF output for the mixer

The noise performance of a mixer circuit is usually characterized by its *noise figure* [1], which can be defined by

$$NF = \frac{total\ IF\ output\ noise\ power}{that\ part\ of\ output\ noise\ power\ due\ to\ source\ resistance\ at\ RF\ port}. \quad (5.149)$$

This definition is intended for circuits in small-signal operation. For such circuits, noise power is a constant function of time. In our case, the noise variance, i.e. the mean-squared noise power, at the output of the mixer circuit changes as a function of time. Thus, we can say that the noise figure is also an instantaneous quantity that varies with time. Hence, we define a *time-varying* noise figure as follows:

$$NF(t) = \frac{total\ IF\ output\ noise\ variance}{IF\ output\ noise\ variance\ due\ to\ source\ resistance\ at\ RF\ port}. \quad (5.150)$$

To calculate the time-varying noise figure as defined, we simulate the mixer circuit again to calculate the noise variance waveform at the output with all of the noise sources turned off except for the noise source for the source resistance $RS_{RF} = 50\ \Omega$ at the RF port. In this case, we obtain the noise variance waveform in Figure 5.7. We now can compute the time-varying noise figure in

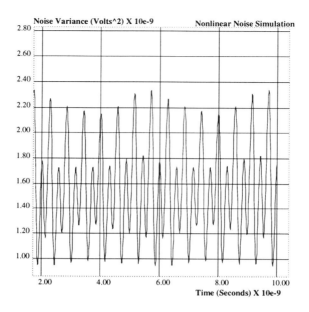

Figure 5.7. Noise voltage variance at the IF output due to $RS_{RF} = 50\ \Omega$

(5.150). The noise figure "waveform" (in dBs) is shown in Figure 5.8. In practice, what one measures with a piece of equipment is usually a single number to characterize the noise figure, because one does not measure the instantaneous noise power as a time-varying quantity. Instead, one measures a *time-average* of the time-varying noise power. Thus, we define the *average* noise figure as follows:

$$NF_{avg} = \frac{avg.\ of\ total\ IF\ noise\ variance}{avg.\ of\ IF\ noise\ variance\ due\ to\ source\ resistance\ at\ RF\ port}. \tag{5.151}$$

We calculate the average noise figure for the mixer by first computing the average noise variances in Figure 5.6 and Figure 5.7, and then by computing the ratio. The result in dBs is $NF_{avg} = 17.9\ dB$.

This bipolar mixer circuit has 65 nodes (including the internal nodes for BJTs) which have capacitive paths to ground. There are a total of 91 noise sources associated with the bipolar transistors and the resistors in the circuit. The numerical solution of (5.87) to calculate the variance-covariance matrix as a function of time (with 400 time points) took 140 CPU seconds on a DEC Alpha machine[8] with our current implementation (with the Bartels-Stewart algorithm used to solve the algebraic Lyapunov matrix equation). In this simulation, 2145

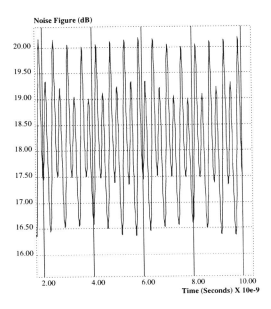

Figure 5.8. Time-varying noise figure for the mixer

noise variance-covariance matrix entries for the 65 nodes are calculated at 400 time points.

We can also compute the correlation function $R_{V_n}(t_{ref}, t) = \mathbf{E}\left[V_n(t_{ref})V_n(t)\right]$ for the noise voltage at the IF output by solving (5.99). Since the noise voltage is a nonstationary process, $R_{V_n}(t_{ref}, t)$ is also a function of a reference time point t_{ref}. For the simulation shown in Figure 5.9, the reference time point was chosen as $t_{ref} = 10 \ nsecs$. The initial condition $R_{V_n}(t_{ref}, t_{ref})$ is obtained from the simulation in Figure 5.6. The correlation function $R_{V_n}(t_{ref}, t)$ for the CMOS inverter satisfies

$$R_{V_n}(t_{ref}, t) \to 0 \quad \text{as} \quad (t - t_{ref}) \to \pm \infty \tag{5.152}$$

because it is also a "stable" circuit.

5.10.4 Negative resistance oscillator

The "negative resistance" oscillator consists of a two-terminal nonlinear voltage-controlled resistor (VCR) with a negative resistance region that is connected across a parallel RLC circuit. For simplicity, the values of the components for

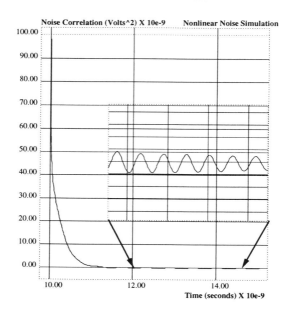

Figure 5.9. Noise voltage correlation for the mixer

the circuit were chosen as
$$R = 1$$
$$C = 1 \qquad (5.153)$$
$$L = 1$$

and the $I - V$ relationship for the VCR is given by

$$I = \tanh(-5\,V). \qquad (5.154)$$

In steady-state, this autonomous circuit settles into a "stable" limit cycle which is illustrated in Figure 5.10. This is a plot of the large-signal capacitor voltage (the voltage across the parallel RLC circuit) versus the large-signal inductor current with time as a parameter. The waveform for the capacitor voltage is periodic in steady-state, as seen in Figure 5.11.

The thermal noise source for the resistor in the circuit is turned off. Instead, a white Gaussian WSS current noise source with a double-sided spectral density

$$S_n(f) = 10^{-10}\ A^2/Hz \qquad (5.155)$$

is connected across the parallel RLC circuit. (5.87) corresponding to this autonomous circuit was numerically solved to compute the capacitor noise voltage

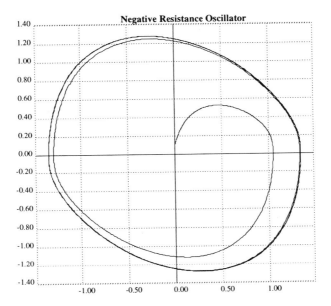

Figure 5.10. Limit cycle for the negative resistance oscillator

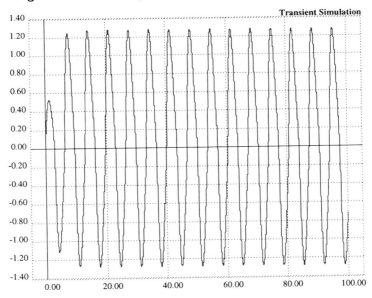

Figure 5.11. Oscillation waveform for the negative resistance oscillator

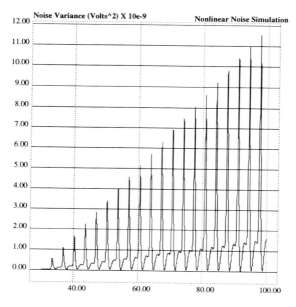

Figure 5.12. Noise voltage variance for the negative resistance oscillator

variance $\mathbf{E}\left[V_n(t)^2\right]$. The initial condition was set to zero as discussed in Section 5.7 and Section 5.9. The waveform calculated for the capacitor noise voltage variance is shown in Figure 5.12. The noise voltage variance does *not* reach a periodic steady-state, it is an oscillatory waveform with a *linear* ramp envelope. (5.87) for this circuit does *not* have a periodic steady-state solution. Thus, the capacitor noise voltage is not a cyclostationary process in steady-state even though the circuit is in large-signal periodic steady-state. The peaks of the voltage variance waveform coincide with the transitions, i.e. zero-crossings, of the large-signal periodic voltage waveform in Figure 5.11.

We then compute the correlation function $R_{V_n}(t_{ref}, t) = \mathbf{E}[V_n(t_{ref})V_n(t)]$ for the capacitor noise voltage by solving (5.99). Since the noise voltage is a nonstationary process, $R_{V_n}(t_{ref}, t)$ is also a function of a reference time point t_{ref}. For the waveform shown in Figure 5.13, the reference time point was chosen as $t_{ref} = 70.7\ secs$ to coincide with one of the peaks in the waveform of Figure 5.12. The initial condition $R_{V_n}(t_{ref}, t_{ref})$ is obtained from the waveform in Figure 5.12. (5.99) for this oscillator circuit has a periodic steady-state solution. The correlation function $R_{V_n}(t_{ref}, t)$ does *not* satisfy

$$R_{V_n}(t_{ref}, t) \to 0 \quad \text{as} \quad (t - t_{ref}) \to \pm\infty. \tag{5.156}$$

TIME-DOMAIN NON-MONTE CARLO NOISE SIMULATION   159

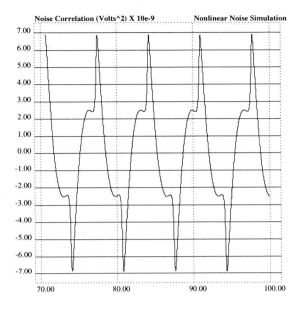

Figure 5.13. Noise voltage correlation for the negative resistance oscillator

We will discuss the interpretation of the noise simulation results for the negative resistance oscillator circuit in Chapter 6.

5.11 Summary

We presented a time-domain non-Monte Carlo noise simulation algorithm for nonlinear dynamic circuits with arbitrary large-signal excitations. The nonlinear network equations (KVL, KCL and the constitutive relations of the components) that govern the behavior of a nonlinear circuit with noise sources were formulated as a system of stochastic differential equations. Then, we set out to calculate a probabilistic characterization of the circuit variables (e.g. node voltages) that are modeled with stochastic processes. We discussed the practical infeasibility of the calculation of a complete probabilistic characterization that requires the solution of the Fokker-Planck equation for the time evolution of the probability density of the state vector of the nonlinear system. The "small-signal" nature of the noise signals led us to a stochastic small noise expansion of the state vector and the system of nonlinear stochastic differential equations. As a result of the small noise expansion, the state vector of the system was decomposed as a summation of a deterministic large-signal component and a stochastic noise one. A linear but time-varying system of stochastic differential equations was derived which describes the dynamics of the stochastic noise component. The deterministic time-varying coefficient matrices for the system of LTV stochastic differential equations are dependent on the solution of the system of nonlinear ODEs that describe the behavior of the nonlinear circuit without the noise sources. Then, using stochastic calculus, two sets of systems of ODEs for the autocorrelation matrix of the stochastic noise component of the system state vector were derived. Assuming that all of the noise sources are Gaussian, the stochastic noise component of the system state vector is a vector of nonstationary zero-mean *Gaussian* stochastic processes, because the system of stochastic differential equations that describes its dynamics is *linear* although time-varying. Thus, the autocorrelation matrix is a complete probabilistic characterization. We then presented techniques to numerically solve the ODEs to compute the autocorrelation matrix for the stochastic noise component of the system state vector. The conditions for the existence of "valid" solutions for the system of ODEs for the autocorrelation matrix were discussed. Then, two special cases, namely when the nonlinear circuit is in large-signal time-invariant or periodic steady-state were discussed. For these special cases, the conditions for the existence of steady-state solutions of the ODEs for the autocorrelation matrix were presented along with a description of their numerical computation. Finally, we presented several practical examples of the numerical computation of the autocorrelation function of a noise signal in nonlinear circuits. These were simple examples which meant to describe how to use the time-domain non-Monte Carlo noise simulation algorithm to characterize the noise performance of practical nonlinear circuits. In particular, we discussed the switching noise of an inverter that is driven with a periodic waveform, and the computation of the noise figure of a mixer that is driven with two signals at its RF and LO inputs. We also used the noise simulation algorithm on an autonomous nonlinear circuit.

The time-domain non-Monte Carlo noise simulation algorithm, and its implementation in the circuit simulator SPICE, enables us to calculate the complete second-order probabilistic characteristics of the state variables (e.g. node voltages) of a nonlinear circuit under the influence of both large-signal deterministic excitations and noise sources. In this sense, it is a *core* tool which will enable us to *investigate, simulate, understand,* and *model* various phenomena that is related to noise, and which is of concern from a performance point of view, in nonlinear circuit design. As a result, it will also enable us to develop more specific algorithms and numerical techniques to characterize specific noise phenomena, and *define* and compute quantities that will become noise performance measures.

In the next chapter, we will use the core noise simulation algorithm to investigate, understand and model an extremely important phenomenon in oscillator circuits that is related to noise, the so-called *phase noise* or *timing jitter*. As a result of this analysis, we will arrive at a more specific algorithm to characterize phase noise/timing jitter for oscillators. We will develop models which can be used in various ways to analyze the effect of the phase noise/timing jitter of an oscillator on the performance of a larger system that contains the oscillator as a component. This will be a perfect illustration of the enabling features (claimed above) of the core noise simulation technique that was presented.

Notes

1. See [50] for an example of calculating analytical solutions of the Fokker-Planck equation. [50] uses the Fokker-Planck equations to study the "cycle-slipping" behavior of simple phase-locked loops.

2. In an actual numerical calculation, the inverse of $C_{11}(t)$ is not explicitly calculated. Instead, the LU decomposition of $C_{11}(t)$ is followed by a number of forward elimination and backward solution steps to effectively multiply both sides of (5.49) by the inverse of $C_{11}(t)$.

3. See Section 2.4.9 for the definition. Recall that the Floquet exponents are not related to the eigenvalues of $\mathbf{E}(t)$, but they are related to the eigenvalues of the state transition matrix $\mathbf{\Phi}(T,0)$ of (5.85).

4. In the examples to be presented, only the *shot* and *thermal* noise sources associated with the electronic components and the semiconductor devices have been included in the device models.

5. This is obtained from the numerical solution of (5.87).

6. This is obtained from the numerical solution of (5.99).

7. For this case, the noise voltage at the output is cyclostationary at steady-state, because we are driving the inverter with a periodic waveform.

8. The CPU is a DEC Alpha chip 21164 with 250 MHz clock frequency, 4 Mb of cache, and a SPEC int_92 of 277.

6 NOISE IN FREE RUNNING OSCILLATORS

Oscillators are among the key components of many different kinds of electronic systems. They are used for on-chip clock generation for microprocessors. Every communications receiver/transmitter has at least one oscillator that is used in the frequency synthesis of an oscillation signal which up or down converts the incoming/outgoing signal. Oscillators have *one* property that makes them quite unique from several aspects: They are *autonomous* systems. They generate an oscillatory signal at their output without an input (apart from a power supply input, and a control signal that sets the frequency), as opposed to amplifiers and mixers which generate an output when they are being driven with some input signals. The design, analysis and simulation of oscillators often require techniques which are specific for autonomous systems.

An oscillator can be defined as a system that generates a *periodic* signal with a specified or controllable frequency, but any periodic signal with the specified frequency is not acceptable. An oscillator is not just any autonomous system that generates a periodic signal. It is hard to formalize the distinction of a signal an oscillator is supposed to generate from any periodic signal, but ideally, we would like it to be as close as to a *square-wave* or a *pure sinusoid* at the specified frequency. Square-wave-like waveforms with fast transitions (between "high" and "low" states) that are evenly spaced in time are desirable in some applications. In other applications, a sinusoidal waveform is the choice because of its spectral properties.

The autonomous nature of oscillators also makes them quite unique in their behavior in response to the electrical noise that is present in the circuit. The analysis and characterization of oscillators in the presence of noise is quite an intricate topic. Because of the practical importance of understanding and characterizing the behavior of oscillators in the presence of noise, this topic attracted considerable attention in the literature. It is quite impossible to review all the approaches that were proposed to analyze oscillators in the presence of noise.[1] Most of the approaches in the literature for oscillator noise analysis use techniques from the theory of LTI systems and WSS stochastic processes. This is obviously not justified at all, because practical oscillators are always nonlinear systems. Many try to explain the experimental measurements of the "spectrum" of practical oscillator circuits and develop models based on the results of the LTI/WSS analysis. A small number of approaches do use techniques suitable for analysis of nonlinear systems in the presence of noise, but most of these are analytical approaches for specific oscillator circuits. Moreover, they use simplified models of the circuit and make simplifying assumptions, because the analysis of a nonlinear oscillator circuit in the presence of noise is not analytically tractable. Even a smaller number of approaches treat the problem in general, and propose methods that are amenable for a numerical implementation to simulate and characterize the noise performance of oscillators on the computer.

We believe that there is a lot of confusion on the definitions of certain notions connected to oscillator noise analysis. Some of the papers in the literature characterize the noise behavior of oscillators using the notions of *amplitude noise*, *phase noise*, and *timing jitter*. Very few of them fully define exactly what they mean by amplitude and phase noise, and timing jitter. Some other papers do not even discuss the distinction of amplitude and phase noise, and talk about oscillator noise in general. In this chapter, we will first discuss the effect of noise on the periodic signal that is generated by an oscillator, and then make an attempt to clarify the notions of amplitude and phase noise, and timing jitter, and discuss their relevance from a practical point of view. We will give a definition for *phase noise/timing jitter* as a stochastic process, and present an algorithm to characterize these quantities that is based on the time-domain noise simulation technique we presented in the previous chapter. We will prove a key property of our definition of phase noise and the characterization algorithm: Phase noise is same at all nodes in the oscillator circuit. We will then discuss the work of Franz Kaertner [16] on phase noise (a rigorous approach to noise analysis of oscillators), and review his phase noise definition and characterization algorithm and its connection to our definition and characterization algorithm. In this chapter, we will be mainly concerned with the noise analysis of *free running* or *open-loop* oscillators, but we will briefly discuss the noise performance of *closed-loop* oscillators, i.e. phase-locked loops, which will be the main topic of the next chapter.

6.1 Phase Noise and Timing Jitter Concepts

We have already used the time-domain noise simulation technique on an oscillator in Section 5.10.4. With our noise simulation technique, we decompose a "noisy" oscillation signal $X(t)$ as follows:

$$X(t) \approx x_s(t) + X_n(t) \tag{6.1}$$

where $x_s(t)$ is the deterministic oscillation waveform that is periodic with T, and $X_n(t)$ is a stochastic process that represents the effect of the noise sources. We first simulate the circuit to numerically calculate $x_s(t)$. Then, we calculate the autocorrelation function of $X_n(t)$ for a complete second-order probabilistic characterization. We have found out that the variance of $X_n(t)$ ($\mathbf{E}\left[X_n(t)^2\right]$ in Figure 5.12) for the negative resistance oscillator circuit does not reach a periodic steady-state. It is an oscillatory waveform with a *linear* ramp envelope that grows *without bound*. This result, at first sight, is rather counter intuitive. It *suggests* that the oscillation waveform is becoming "noisier" or "fuzzier" as time progresses, and the oscillator is drifting away from a stable oscillation. However we know that this can not be true. The negative resistance oscillator we considered in Section 5.10.4 *does* settle into a stable limit cycle, and the noise in the circuit causes only small fluctuations in the oscillation waveform, which we state here without proof. Then, how do we resolve this dilemma and interpret the results we obtained with time-domain noise simulation? The resolution of this dilemma lies in the following observations: The *peaks* (which have a linear ramp envelope) in the waveform of $\mathbf{E}\left[X_n(t)^2\right]$ coincide (in time) with the transitions of the large-signal periodic waveform for $x_s(t)$, i.e. the zero-crossings in Figure 5.11. The *dips* in $\mathbf{E}\left[X_n(t)^2\right]$ coincide (in time) with the "high" and "low" states of the large-signal periodic waveform for $x_s(t)$, i.e. the peaks and dips in Figure 5.11. If $\mathbf{E}\left[X_n(t)^2\right]$ is sampled at the peaks, we obtain a linear ramp waveform. However, if it is sampled at the dips, then we obtain a waveform that settles to a time-invariant steady-state. Thus, the variance of the noise in the oscillation waveform at the transitions grows without bound, but the variance at the peaks and dips of the oscillation waveform does not grow without bound, moreover it settles to a time-invariant steady-state value.

Next, we consider a sinusoidal waveform to gain more insight into the above observations. Let
$$x_s(t) = A \cos(2\pi f_c t) \qquad (6.2)$$
be a noiseless sinusoidal waveform with frequency f_c and amplitude A. Let us represent a noisy sinusoidal waveform with
$$X(t) = (A + a(t)) \cos(2\pi f_c t + \phi(t)) \qquad (6.3)$$
where $a(t)$ and $\phi(t)$ are possibly *nonstationary* and correlated zero-mean stochastic processes, and will be referred to as the *amplitude* and the *phase* noise respectively. We define the stochastic process $X_n(t)$ to be the difference between the noisy and the noiseless sinusoidal waveform, i.e.
$$X_n(t) = (A + a(t)) \cos(2\pi f_c t + \phi(t)) - A \cos(2\pi f_c t). \qquad (6.4)$$

Now, we would like to assign properties to the stochastic processes $a(t)$ and $\phi(t)$ so that $\mathbf{E}\left[X_n(t)^2\right]$ for the sinusoidal waveform will "look like" the one we calculated for the negative resistance oscillator in Figure 5.12. If the amplitude and phase noise are "small", we can approximate $X_n(t)$ as follows:
$$X_n(t) \approx a(t) \cos(2\pi f_c t) - A\phi(t) \sin(2\pi f_c t). \qquad (6.5)$$

We observe that the first term in (6.5) vanishes at the zero-crossings of $x_s(t) = A \cos(2\pi f_c t)$, and the second term vanishes at the peaks and dips of $x_s(t) = A \cos(2\pi f_c t)$. Thus, we choose $a(t)$ to be a WSS process, and choose $\phi(t)$ to be a scaled version of the Wiener process, so that $\mathbf{E}\left[X_n(t)^2\right]$ for the sinusoidal waveform will "look like" the one we calculated for the negative resistance oscillator in Figure 5.12. Figure 6.1 shows an *ensemble* of noisy sinusoidal waveforms (all at the same frequency f_c) compiled with a WSS $a(t)$ and a Wiener process $\phi(t)$. All of the waveforms in the ensemble are synchronized at $t = 0$. Recall that for a standard Wiener process $W(t)$, we have
$$W(t) = 0 \qquad (6.6)$$
$$\mathbf{E}\left[W(t)^2\right] = t. \qquad (6.7)$$

Figure 6.2 shows the ensemble for $X_n(t)$ obtained by using (6.4) on all of the waveforms in the ensemble of Figure 6.1. Figure 6.2 also shows the noiseless sinusoidal waveform. Then, Figure 6.3 shows the variance $\mathbf{E}\left[X_n(t)^2\right]$ as a function of time, which was calculated by taking expectations over the ensemble in Figure 6.2. Each of the noisy sinusoidal waveforms in the ensemble of Figure 6.1, when plotted by itself, can not be differentiated from the noiseless waveform since the amplitude noise $a(t)$ is WSS and small compared with A. However,

NOISE IN FREE RUNNING OSCILLATORS   167

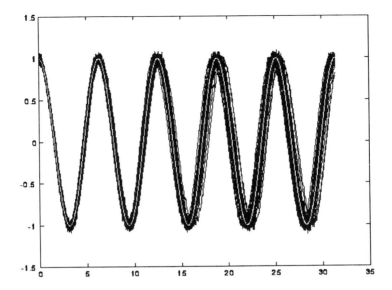

Figure 6.1.   Ensemble of noisy sinusoidal waveforms

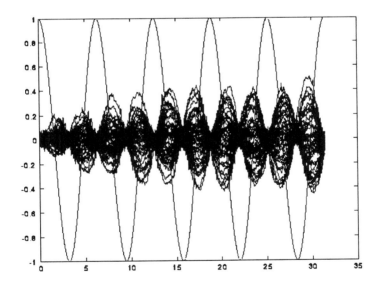

Figure 6.2.   Ensemble of the noise components of noisy sinusoidal waveforms

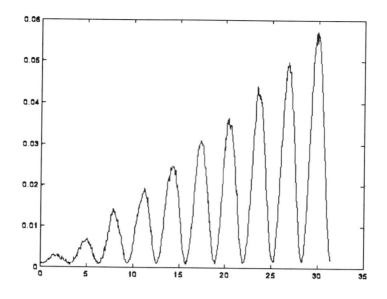

Figure 6.3. Variance of the ensemble for a noisy sinusoidal waveform

due to the nonstationary phase noise $\phi(t)$ with an increasing variance, the error in the zero-crossing times of a noisy sinusoidal waveform with respect to the zero-crossings of the noiseless waveform *increases* as time progresses. A Wiener process $\phi(t)$ can also be interpreted as a *white noise* deviation in the frequency of the sinusoidal waveform. One can also interpret the phase noise $\phi(t)$ as noise in *time*. If we were to *redefine* time t as a stochastic process t' as follows

$$t' = t + \frac{\phi(t)}{2\pi f_c} \tag{6.8}$$

then a sinusoidal waveform with phase noise becomes

$$A\cos\left(2\pi f_c t + \phi(t)\right) = A\cos\left(2\pi f_c t'\right). \tag{6.9}$$

Thus, phase noise is equivalent to a time or *timing* noise for the oscillatory waveform which manifests itself as jitter in the zero-crossing times.

In most applications, the jitter in the timings of the zero-crossings of oscillation waveforms due to noise is of practical importance. The noise in the amplitude of the signal is usually "cleaned" by passing it through a limiter, and it is unimportant in many applications, for instance, when the oscillator output is fed into a digital phase detector or a switching mixer. These circuits are sensitive to only the jitter in the timing of the transitions, and not to the noise in the amplitude of the signal. We would like to model and characterize the timing jitter, i.e. phase noise, as a stochastic process. In particular, we would like to calculate its *complete* second-order probabilistic characteristics, i.e. the autocorrelation function or the (possibly time-varying) spectral density. The "spectrum" of a noiseless oscillation waveform consists of impulses at the frequency of oscillation and its harmonics, where *spectrum* is defined as the Fourier transform of the periodic deterministic oscillation waveform. If we model a noisy oscillation waveform as a stochastic process, then a "spectrum" may be defined through spectral densities of stochastic processes. The notion of a "spectrum" for a noisy oscillation waveform has been used extensively in the literature, almost everywhere without an exact definition. Gardner in [50] discusses the problems in characterizing the noise of an oscillator with a "spectrum".

In the next section, we will formalize the notion of *timing jitter* or *phase noise* which both refer to the same phenomenon, i.e. the noise in the zero-crossing or transition times of oscillation waveforms. We will first present a definition of phase noise as a discrete-time stochastic process, and then describe an algorithm for its characterization along with examples. The above discussion of amplitude and phase noise for a sinusoidal waveform was meant to give some intuition into the notion of phase noise or timing jitter, and provide motivation for the definition we will present next.

6.2 Phase Noise Characterization with Time Domain Noise Simulation

We assume that the behavior of the autonomous oscillator circuit (without the noise sources) is governed by the following system of equations in MNA form

$$\mathbf{I}(\mathbf{x}) + \frac{d}{dt}\mathbf{Q}(\mathbf{x}) = 0 \qquad (6.10)$$

which was obtained from (5.3) by omitting the explicit time dependence of $\mathbf{I}$, because the autonomous oscillator circuit does not have any time-varying deterministic excitations.

Assumption 6.1 *We assume that (6.10) has a periodic steady-state solution* $\mathbf{x}_s(t)$, *which is a stable[2] of limit cycle in the n-dimensional space.*

Thus, $\mathbf{x}_s(t)$ satisfies

$$\mathbf{x}_s(t) = \mathbf{x}_s(t + kT) \qquad (6.11)$$

for all $t, k \in \mathbf{Z}$ and for some period T. We define $f_c = 1/T$ to be the frequency of the periodic $\mathbf{x}_s(t)$. The equations for the oscillator circuit with the noise sources is formulated as follows

$$\mathbf{I}(\mathbf{X})dt + d\mathbf{Q}(\mathbf{X}) + \mathbf{B}(\mathbf{X})d\mathbf{W}(t) = 0 \tag{6.12}$$

as a system of *stochastic* algebraic and differential equations, which was obtained from (6.12) by omitting the explicit time dependence of $\mathbf{I}$ and $\mathbf{B}$. With the small noise expansion described in Section 5.3, we approximate the solution of (6.12) with

$$\mathbf{X}(t) \approx \mathbf{x}_s(t) + \mathbf{X}_n(t). \tag{6.13}$$

With the time domain noise simulation algorithm described in Chapter 5, we can calculate the autocorrelation function of the components of the vector of stochastic processes $\mathbf{X}_n$, which represent the noise component of the node voltages for the nodes which have a capacitive path to ground and the noise component of the inductor currents.

6.2.1 Definition of timing jitter and phase noise

We will now formally define timing jitter and phase noise for the oscillator circuit. We will define timing jitter and phase noise for a specific node (which has a capacitive path to ground) voltage or an inductor current. Whether it is a node voltage or an inductor current, this circuit variable is a component of the state vector $\mathbf{x}$ for the circuit. Let this circuit variable be the ith component of the state vector. Hence

$$x_s(t) = \{\mathbf{x}_s(t)\}_i \tag{6.14}$$
$$X_n(t) = \{\mathbf{X}_n(t)\}_i \tag{6.15}$$

where $\{.\}_i$ denotes the ith component of a vector. Thus, $x_s(t)$ is the deterministic periodic steady-state waveform for the circuit variable, and $X_n(t)$ is the noise component. Let

$$S(t) = \frac{d}{dt}x_s(t) \tag{6.16}$$

be the time derivative of the periodic steady-state waveform $x_s(t)$. Hence, $S(t)$ is also periodic with the same period T. Let us now define the set Γ of evenly spaced time points as

$$\Gamma = \{\tau_k \geq 0 : S(\tau_k) = \max_{0 \leq t \leq T} S(t),\ \tau_k - \tau_{k-1} = T\}. \tag{6.17}$$

Thus, Γ is the set of time points where the periodic oscillation waveform $x_s(t)$ is making low-to-high transitions. The definition of Γ as given above makes sense only for a certain class of periodic waveforms. For instance, it does not make sense for a triangle-wave $x_s(t)$, because $S(t)$ for a triangle-wave is a periodic piecewise constant function, and hence $x_s(t)$ does not have well-defined low-to-high transition times that can be identified as the time points where $x_s(t)$ has the highest *slew rate*, i.e. the time-derivative or the rate of change. However, it is well-defined for a sinusoidal waveform, or a square-wave with a finite slope during the short switching times. It is hard to give a formal characterization of periodic waveforms for which Γ is well-defined, but one can roughly say that Γ is well-defined for periodic waveforms which have a high and a low "state" and transitions between these that are identifiable as the time points with highest time-derivative, i.e. waveforms we would like the oscillators to produce. One property of these periodic waveforms is that the periodic waveforms obtained as their time derivatives *look like* themselves. For instance, the derivative of a sinusoidal waveform is also a sinusoidal waveform, and the derivative of a square-wave with a finite slope during the transitions is also square-wave (though with a duty cycle that is smaller then 50%). It is very plausible that every practical oscillator circuit will have a circuit variable for which Γ will be well-defined. This will be the case for all of the practical oscillator circuits we will consider. Note that, for the definition of Γ, we have arbitrarily chosen the low-to-high transitions times. As we will see in the next section, choosing the high-to-low transitions yields exactly the same results for phase noise or timing jitter characterization. Let S be the maximum value of the derivative $S(t) = \dot{x}_s(t)$, i.e.

$$S = \max_{0 \leq t \leq T} S(t). \tag{6.18}$$

Note that
$$S(\tau_k) = S \quad \text{for} \quad \text{all} \quad \tau_k \in \Gamma \tag{6.19}$$

We define *timing jitter* to be the discrete-time stochastic process J

$$J[k] = \frac{X_n(\tau_k)}{S(\tau_k)} = \frac{X_n(\tau_k)}{S} \tag{6.20}$$

for $\tau_k \in \Gamma$ and $k \geq 0$. Thus, the timing jitter J is a sampled and scaled version of $X_n(t)$. Note that J has the units of time. Then, *phase noise* is defined as the discrete-time stochastic process ϕ

$$\phi[k] = 2\pi f_c \frac{X_n(\tau_k)}{S} \tag{6.21}$$

which is only a scaled version of the timing jitter J. ϕ is in *radians*. Thus, timing jitter and phase noise are basically the same discrete-time stochastic process up to a scaling factor that is equal to the angular frequency $2\pi f_c$.

If we use the definition of phase noise in (6.21) on the noisy sinusoidal waveform discussed in the previous section (see (6.2) and (6.5)), we obtain a sample of the continuous-time phase noise process $\phi(t)$ that was introduced to the noiseless sinusoidal waveform. By sampling $X_n(t)$ in (6.5) at the zero-crossings or transition times of $x_s(t)$ given in (6.2), we *reject* the first term in (6.5) that is due to amplitude noise and obtain a sample of the continuous-time phase noise process $\phi(t)$. Thus, the definition of phase noise given by (6.21) is consistent with the usual definition of *phase* for a sinusoidal waveform. For nonsinusoidal oscillation waveforms, there is no obvious definition for phase, but we still use the terminology *phase* noise to describe the noise in the zero-crossing times of general oscillation waveforms as defined by (6.21).

6.2.2 Probabilistic characterization of phase noise

Now, we would like to calculate a second-order probabilistic characterization of phase noise and timing jitter as defined. Using the time-domain noise simulation technique described in Chapter 5, we can numerically calculate the autocorrelation function of $X_n(t)$ for the circuit variable we are considering. Since phase noise is obtained by sampling the stochastic process $X_n(t)$, its autocorrelation function can be easily calculated by sampling the autocorrelation function of $X_n(t)$. $\mathrm{E}\left[X_n(t)^2\right]$ as a function of t is obtained by numerically solving (5.75), and $\mathrm{E}\left[X_n(t_{ref})X_n(t)\right]$ as a function t for a reference time point $t_{ref} \leq t$ is obtained by numerically solving (5.81). Then, for phase noise ϕ, we have

$$\mathrm{E}\left[\phi[k]^2\right] = \beta^2 \mathrm{E}\left[X_n(\tau_k)^2\right] \tag{6.22}$$
$$\mathrm{E}\left[\phi[k_{ref}]\phi[k]\right] = \beta^2 \mathrm{E}\left[X_n(\tau_{k_{ref}})X_n(\tau_k)\right] \tag{6.23}$$

where $k \geq k_{ref}$, and $\tau_{k_{ref}}, \tau_k \in \Gamma$, and

$$\beta = \frac{2\pi f_c}{S}. \tag{6.24}$$

S was defined by (6.18). For timing jitter,

$$\mathrm{E}\left[J[k]^2\right] = \frac{\mathrm{E}\left[\phi[k]^2\right]}{(2\pi f_c)^2} \tag{6.25}$$
$$\mathrm{E}\left[J[k_{ref}]J[k]\right] = \frac{\mathrm{E}\left[\phi[k_{ref}]\phi[k]\right]}{(2\pi f_c)^2} \tag{6.26}$$

follow from (6.22) and (6.23).

Let us now calculate the second-order probabilistic characterization of phase noise for the negative resistance oscillator described in Section 5.10.4. We choose the capacitor voltage as the circuit variable for phase noise characterization. Figure 5.11 shows the steady-state deterministic oscillation waveform

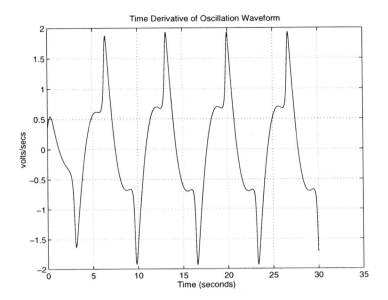

Figure 6.4. Time-derivative of the oscillation waveform for the negative resistance oscillator

$x_s(t)$. The waveform that is obtained as the time derivative of $x_s(t)$ in Figure 5.11, i.e. $\dot{x}_s(t)$, is shown in Figure 6.4, from which we identify Γ as defined by (6.17), and also $S = 1.93$ volts/secs that was defined in (6.18). Notice that the waveform of $\dot{x}_s(t)$ looks like exactly the waveform in Figure 5.13. We will see the reason for this in Section 6.3. The waveform obtained for the noise variance $\mathbf{E}\left[X_n(t)^2\right]$ of the capacitor voltage is in Figure 5.12. This is an oscillatory waveform with a *linear* ramp envelope that grows *without bound*. The *peaks* in the waveform of $\mathbf{E}\left[X_n(t)^2\right]$ coincide (in time) with the transitions of the large-signal periodic waveform for $x_s(t)$, i.e. the time points in Γ. Hence, if $\mathbf{E}\left[X_n(t)^2\right]$ is sampled to calculate $\mathbf{E}\left[\phi[k]^2\right]$ as defined by (6.22) for the discrete-time stochastic process ϕ that represents phase noise, we obtain

$$\mathbf{E}\left[\phi[k]^2\right] = \beta^2 \alpha kT \quad k \geq 0 \tag{6.27}$$

where α in volts^2/secs is the slope of the linear ramp envelope for $\mathbf{E}\left[X_n(t)^2\right]$. The waveform obtained for $\mathbf{E}\left[X_n(t_{ref})X_n(t)\right]$ is in Figure 5.13. The reference time point was chosen as $t_{ref} = 70.7\ secs$ to coincide with one of the peaks in the waveform of $\mathbf{E}\left[X_n(t)^2\right]$, i.e. one of the time points in Γ. $\mathbf{E}\left[X_n(t_{ref})X_n(t)\right]$ is a periodic waveform. Similarly, the peaks in $\mathbf{E}\left[X_n(t_{ref})X_n(t)\right]$ coincide with the time points in Γ. Hence, if it is sampled to calculate $\mathbf{E}\left[\phi[k_{ref}]\phi[k]\right]$ as defined by (6.23) for the phase noise process ϕ, we obtain

$$\mathbf{E}\left[\phi[k_{ref}]\phi[k]\right] = \mathbf{E}\left[\phi[k_{ref}]^2\right] \qquad (6.28)$$
$$= \beta^2 \alpha k_{ref} T \qquad (6.29)$$
$$= \beta^2 \mathbf{E}\left[X_n(t_{ref})^2\right] \qquad (6.30)$$

where $t_{ref} = k_{ref}T$ and $k \geq k_{ref}$. It can be shown that (6.29) is valid for all $t_{ref} = k_{ref}T$, not just the simulation results are shown for. Thus, if we combine (6.27) and (6.29), we obtain

$$\mathbf{E}\left[\phi[k]\phi[m]\right] = \beta^2 \alpha \min(k, m)T \quad \text{for} \quad k, m \geq 0. \qquad (6.31)$$

The autocorrelation function given in (6.31) is a complete second-order probabilistic characterization of the discrete-time *zero-mean* stochastic process ϕ that represents the phase noise.

Remark 6.1 *The autocorrelation function given in (6.31) can be identified as the autocorrelation function of a so-called discrete-time* random walk *process R which is constructed as follows:*

$$R[k] = \sum_{r=1}^{k} Z_r \qquad (6.32)$$

where $Z_1, Z_2, \ldots$ are uncorrelated identically distributed zero-mean random variables taking values in $\mathbb{R}$.

Proof. The autocorrelation function of R is given by

$$\mathbf{E}\left[R[k]R[m]\right] = \sum_{r=1}^{k}\sum_{l=1}^{m} \mathbf{E}\left[Z_r Z_l\right] \qquad (6.33)$$
$$= \sum_{r=1}^{\min(k,m)} \mathbf{E}\left[Z_r^2\right] \qquad (6.34)$$
$$= \sum_{r=1}^{\min(k,m)} \sigma^2 \qquad (6.35)$$
$$= \sigma^2 \min(k, m) \qquad (6.36)$$

where

$$\sigma^2 = \mathbf{E}\left[Z_1^2\right] = \mathbf{E}\left[Z_2^2\right] = \cdots \qquad (6.37)$$

□

We obtained the phase noise process ϕ by sampling $X_n(t)$ which is a Gaussian process as explained at the end of Section 5.4. Hence, ϕ is also a Gaussian process. If we let $Z_1, Z_2, \ldots$ be uncorrelated zero-mean *Gaussian* random variables with variance

$$\mathbb{E}\left[Z_1^2\right] = \sigma^2 = \beta^2 \alpha T \tag{6.38}$$

then, the random walk process R will also be Gaussian, and it will have the same autocorrelation function as the phase noise process ϕ. Since the autocorrelation function is a complete characterization of a Gaussian process, then R and ϕ are stochastically equivalent. The representation of the sampled phase noise process as a discrete-time random walk process with uncorrelated, identically distributed, Gaussian increments will be very useful when we discuss behavioral modeling and simulation of phase noise in phase-locked loops in the next chapter. The random walk representation of phase noise allows us to define a *figure of merit* to characterize the phase noise performance of the oscillator. If we have two oscillators at the same frequency, then the one with a larger value of $\sigma^2 = \beta^2 \alpha T$ for phase noise will be worse than the other one from a phase noise performance perspective. $\sigma^2 = \beta^2 \alpha T$ is the rate of change of the variance of the discrete-time phase noise process ϕ.

The random walk phase noise representation we discussed above was derived for the negative resistance oscillator based on the characterization we obtained using time-domain noise simulation. At this point, we do not know that the random walk representation for phase noise is valid for other oscillator circuits. However, we have an algorithm which allows us to numerically calculate the autocorrelation function of the phase noise of the node voltages or the inductor currents of an oscillator:

Algorithm 6.1 (Phase Noise/Timing Jitter Characterization)

1. *We first calculate the steady-state solution of (6.10). This could be done using the transient analysis in SPICE, or with a specific numerical algorithm such as the shooting method for finding periodic steady-state solutions of nonlinear autonomous circuits.*

2. *We identify the frequency of oscillation $f_c = 1/T$.*

3. *We apply the time-domain noise simulation algorithm described in Chapter 5 to calculate the noise autocorrelation functions of the node voltages with a capacitive path to ground and the inductor currents.*

4. *We choose a circuit variable which could be a node voltage, a differential voltage or an inductor current. We choose a node with a steady-state oscillation waveform for which Γ can be defined as discussed before.*

5. We identify Γ and calculate S defined by (6.18).

6. We sample and scale the autocorrelation function (which was calculated numerically with the time-domain noise simulation algorithm) of the circuit variable we have chosen to calculate the autocorrelation function of the discrete-time phase noise process defined by (6.21).

In the next section, we will apply the above algorithm to characterize the phase noise of several practical oscillator circuits.

6.2.3 Examples

We will apply the phase noise characterization methodology to three oscillators: a ring-oscillator, a relaxation oscillator and a harmonic oscillator. For the characterizations to be presented, we consider only the *thermal* and *shot* noise sources of the devices, i.e. the noise sources that can be modeled as *white* Gaussian processes. We will consider the non-white noise sources in Section 6.6.

6.2.3.1 Ring-oscillator.
The ring-oscillator circuit is a three stage oscillator with fully differential ECL buffer delay cells (differential pairs followed by emitter followers). This circuit is from [38]. [38] and [51] use analytical techniques to to characterize the timing jitter/phase noise performance of ring-oscillators with ECL type delay cells. [38] does the analysis for a bipolar ring-oscillator, and [51] does it for a CMOS one. Since they use analytical techniques, they use a simplified model of the circuit and make several approximations in their analysis. [38] and [51] use time-domain Monte Carlo noise simulations to verify the results of their analytical results. They obtain qualitative and some quantitative results, and offer guidelines for the design of low phase noise ring-oscillators with ECL type delay cells. However, their results are only valid for their specific oscillator circuits. We will compare their results with the results we will obtain for the above ring-oscillator using the general phase noise characterization methodology we have proposed which makes it possible to analyze a complicated oscillator circuit without simplifications.

For a ring-oscillator, the natural choice for the circuit variable for phase noise characterization is any tap voltage in the ring. Since the circuit is differential, we use a differential tap voltage for phase noise characterization. Figure 6.5 shows the steady-state deterministic oscillation waveform $x_s(t)$. The noise variance $\mathsf{E}\left[X_n(t)^2\right]$ for the differential voltage is in Figure 6.6, and $\mathsf{E}\left[X_n(t_{ref})X_n(t)\right]$ for $t_{ref} \approx 50$ nsecs is in Figure 6.7. The waveform obtained for the variance in Figure 6.6 is an oscillatory waveform with a *linear ramp* envelope, The peaks of this oscillatory waveform coincide with the transition

NOISE IN FREE RUNNING OSCILLATORS 177

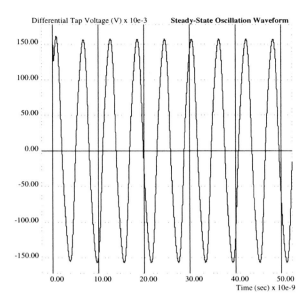

Figure 6.5. Large-signal oscillation waveform for the ring-oscillator

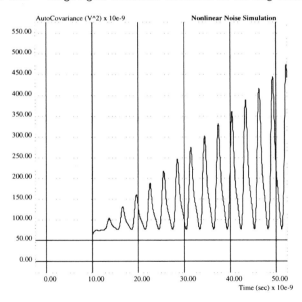

Figure 6.6. Differential noise voltage variance for the ring-oscillator

178   NOISE IN NONLINEAR ELECTRONIC CIRCUITS

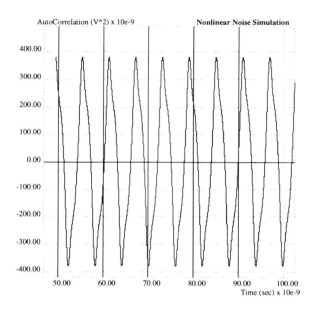

Figure 6.7.   Differential noise voltage correlation for the ring-oscillator

times of $x_s(t)$, i.e. the time-points in Γ. The waveform for $\mathrm{E}\left[X_n(t_{ref})X_n(t)\right]$ is a periodic waveform with the peaks coinciding in time with the time points in Γ. The situation is exactly the same as the one for the negative resistance oscillator. Hence, the autocorrelation function for the sampled phase noise process for the differential tap voltage of this oscillator is exactly in the form given by (6.31). Thus, the random walk model for phase noise is also valid for this ring-oscillator. This result is in agreement with the results and experimental observations presented in [38] for this ring-oscillator.

For further comparison of phase noise characterization results obtained by our numerical method and the ones presented in [38], we performed several other phase noise characterizations for the bipolar ring-oscillator. The results are shown in Table 6.1, where R_c is the collector load resistance for the differential pair (DP) in the delay cell, r_b is the zero bias base resistance for the BJTs in the DP, I_{EE} is the tail bias current for the DP, and f_c is the oscillation frequency for the three stage ring-oscillator.

For all of the cases listed in Table 6.1, the noise variance for the differential voltage obtained by time-domain noise simulation had a linear ramp envelope. The random walk phase noise model is valid for all the cases. Hence, the phase noise performance can be characterized by the slope of this ramp envelope α

R_c (Ω)	r_b (Ω)	I_{EE} (μA)	S (volts/μsecs)	α (volts^2/secs)	f_c (MHz)	$\frac{\alpha}{S^2}$ (sec^2·Hz)
500	58	331	186	9.3	167.7	0.269
2000	58	331	287	12.2	74	0.149
500	1650	331	159	17.38	94.6	0.686
500	58	450	259	12.32	169.5	0.182
500	58	600	330	16.40	169.7	0.151
500	58	715	373	19.77	167.7	0.142

Table 6.1. Ring-oscillator phase noise characterization results

together with β that was defined by (6.24). Note that the changes in R_c and r_b affect the oscillation frequency, unlike the changes in I_{EE}. Figure 6.8 shows a plot of $\beta^2 \alpha$ versus I_{EE} using the data from Table 6.1. This prediction of the dependence of phase noise/timing jitter performance on the tail bias current is in agreement with the analysis and experimental results presented in [38] and [51] for ring-oscillators with ECL type delay cells. Note that larger values of $\beta^2 \alpha$ means *worse* phase noise performance.

6.2.3.2 Relaxation oscillator. The relaxation oscillator is a VCO that is based on the emitter-coupled multivibrator circuit [1]. [52] analyzes the "process of jitter production" for this circuit by describing the circuit behavior with a single first-order stochastic differential equation based on a simplified model for the circuit, and lumping all of the noise sources into a single stationary current noise source. [52] arrives at intuitive qualitative results for low jitter relaxation oscillator design. A relaxation oscillator operates in a highly nonlinear fashion due to regenerative switchings. The analysis of the "process of jitter production" is not analytically tractable without reverting to simplifications. With our general phase noise characterization methodology, we can simulate the process of jitter production in a relaxation oscillator numerically without simplifying the circuit.

Figure 6.9 shows the noise variance for the voltage across the timing capacitor. The envelope for the variance waveform is also a linear ramp as it was the case for the negative resistance and also the ring-oscillator. Obviously, the spikes in the variance waveform coincide with the regenerative switchings of the

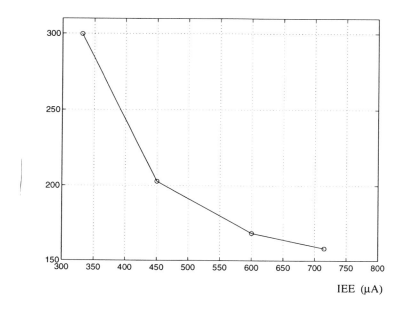

Figure 6.8. Phase noise performance versus I_{EE} for the ring-oscillator

multivibrator. We conclude that the random walk phase noise model is also valid for this relaxation oscillator. The large signal waveform for the timing capacitor voltage is triangular. Hence, switching times characterized by largest slope are not well defined, i.e. Γ for the voltage across the timing capacitor can not be constructed. For phase noise characterization, we choose the output of the multivibrator, which is a square-wave-like waveform. Going through the steps of the phase noise characterization algorithm, we obtain

$$f_c = 0.88 \, \text{MHz} \tag{6.39}$$
$$\alpha = 1.3 \times 10^5 \, \text{volts}^2/\text{sec} \tag{6.40}$$
$$S = 3.29 \times 10^9 \, \text{volts}/\text{sec} \tag{6.41}$$

which results in

$$\beta^2 \alpha = 0.37 \, \text{rad}^2.\text{Hz}. \tag{6.42}$$

6.2.3.3 Harmonic oscillator. The harmonic oscillator has an LC tank, several inductors and a single bipolar-junction transistor with a Colpitts feedback circuit around it. The oscillation frequency is 773.2 MHz. The large-signal

NOISE IN FREE RUNNING OSCILLATORS 181

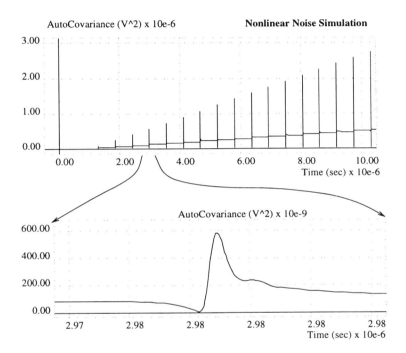

Figure 6.9. Timing capacitor voltage noise variance for the multivibrator

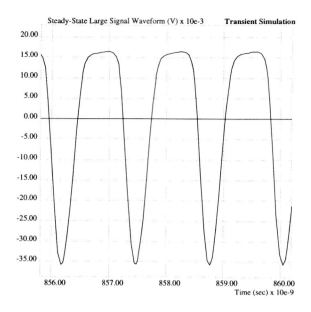

Figure 6.10. Oscillation waveform for the harmonic oscillator

Node	S (volts/nsecs)	α (volts^2/secs)	$\beta^2 \alpha$ (rad^2.Hz)
1	1.53	0.127	1.28
2	1.104	0.063	1.22
3	1.69	0.151	1.25

Table 6.2. Harmonic oscillator phase noise characterization with different nodes

oscillation waveform at the output is shown in Figure 6.10. Time-domain noise simulation showed that the random walk phase noise model is also valid for this oscillator. Hence, phase noise performance can be summarized with the parameters α and β defined before. We calculated $\beta^2 \alpha$ for three different nodes in the circuit to compare the phase noise characterizations for different nodes. The results are in Table 6.2. We can observe in Table 6.2 that even though the large signal slew rate at the transitions, i.e. S, and the slope of the linear ramp envelope for the noise variance waveform, i.e. α, have different values, the calculated value of $\beta^2\alpha$ is equal for the three nodes considered. Hence, *stochastically equivalent* discrete-time random walk processes are obtained for all of the nodes which were considered for phase noise characterization. This intuitively makes sense. Phase noise or timing jitter is basically equivalent to noise in the zero-crossing or transition times of the oscillator waveforms, and we expect it to be equal in all of the waveforms of an oscillator circuit. In Section 6.3, we will show that the phase noise characterization obtained with the definition and the characterization algorithm we have presented is, in general, same for all of the circuit variables of an oscillator. Thus, it is a *property* of the oscillator itself, not just a property of one of the state variables.

The ring-oscillator and the relaxation oscillator we have considered are fully differential circuits. Hence, the low-to-high and high-to-low transitions of the large signal waveforms were symmetric. So, the definition of phase noise obtained as discrete-time stochastic process by sampling at either kind of transition is equivalent. On the other hand, the harmonic oscillator we are considering is not a fully differential circuit, and the low-to-high and high-to-low transitions are not symmetric (i.e. the largest slew rate value is different). However, the phase noise characterization obtained by sampling at either kind of transition for the same node gave the same result, similar to the case where results were compared by calculating the phase noise characterization for different nodes. In Section 6.3, we will also show that this is true in general for the definition of phase noise we presented.

6.2.3.4 Conclusions. We defined phase noise as a discrete-time stochastic process, and found out that it can be characterized as a random walk process for several practical free running oscillators for which the oscillation mechanisms are quite different. Our definition of phase noise was for a specific circuit variable of the oscillator. However, we observed in our phase noise characterization of the harmonic oscillator that, we indeed obtain the same characterization for different circuit variables of the same oscillator. In Section 6.3, we will take a mathematical viewpoint and explore the reason behind these findings, and show that for a class of oscillators (with white noise sources only, i.e. thermal and shot noise) the random walk phase noise model is valid in general, and the phase noise characterization obtained is independent of the circuit variable used.

6.2.4 Phase noise "spectrum"

Almost all of the work in literature on phase noise characterizes it in frequency domain with a "spectrum". Phase noise is clearly a *nonstationary* stochastic process as we have found out for the several practical oscillator circuits we have considered above. It is neither WSS nor cyclostationary, its variance is a monotonically increasing function of time. It is not clear how one would define and characterize a "spectrum" for phase noise.

In our above discussion, we defined phase noise as a discrete-time stochastic process, and gave a full second-order probabilistic characterization for it as a random walk process. We believe that this is the most suitable characterization, given that it is a process without a steady-state characteristics. However, to show the connection of our characterization and the frequency domain characterization usually used for phase noise, we will next discuss the issues that arise when a frequency domain spectrum is derived for phase noise.

We have defined phase noise as a *discrete-time* stochastic process to characterize the noise in the transition times of an oscillation waveform. For frequency domain characterization, we will consider a continuous-time "equivalent" for it. Thus, we define a continuous-time phase noise process $\psi(t)$ which satisfies

$$\psi(kT) = \phi[k] \tag{6.43}$$

where ϕ is the discrete-time phase noise process we have defined earlier. The autocorrelation function of ϕ was given in (6.31). Then, the autocorrelation of ψ, i.e.

$$\mathrm{E}\left[\psi(t+\tau/2)\psi(t-\tau/2)\right] \tag{6.44}$$

should satisfy

$$\mathrm{E}\left[\psi(kT)\psi(mT)\right] = \beta^2 \alpha \min\left(kT, mT\right) \quad \text{for} \quad k, m \geq 0 \tag{6.45}$$

when $t + \tau/2 = kT$ and $t - \tau/2 = mT$ for $t \geq 0$ and $-2t \leq \tau \leq 2t$. Obviously, (6.43) and (6.45) only partially characterize ψ as a stochastic process. As long as (6.43) and (6.45) are satisfied, ϕ and the sampled ψ are stochastically equivalent, and the probabilistic characteristics of ψ is only relevant at the transition times. Hence, we use the extra freedom to specify ψ as a Gaussian stochastic process with the autocorrelation function

$$\begin{aligned} R_\psi(t, \tau) &= \mathrm{E}\left[\psi(t+\tau/2)\psi(t-\tau/2)\right] & (6.46) \\ &= \beta^2 \alpha \min\left(t+\tau/2, t-\tau/2\right) & (6.47) \end{aligned}$$

for $t \geq 0$ and $|\tau| \leq 2t$. Recall that ϕ is also a Gaussian process. The autocorrelation function in (6.47) obviously satisfies (6.45), and it is the autocorrelation function of a (scaled) Wiener process. The scaled Wiener process ψ is the output of an ideal integrator when the input is a white noise source with spectral density

$$S_i(f) = \beta^2 \alpha. \tag{6.48}$$

The transfer function (as defined by (2.107)) of an ideal integrator is given by

$$H_{int}(f) = \frac{1}{2}\delta(f) + \frac{1}{j2\pi f}. \qquad (6.49)$$

If we use (6.48), (6.49) and (2.131) to calculate the "spectral density" of the output of the integrator, we obtain

$$S_\psi(f) = \beta^2 \alpha \frac{1}{4\pi^2 f^2} \qquad (6.50)$$

by ignoring the singularitites at $f = 0$. The spectrum expression above is the one that is extensively used to characterize the phase noise of free running oscillators. In the phase noise literature, it is most often interpreted as if it is the spectral density of a WSS process that models the phase noise of the oscillator. This interpretation is clearly wrong. The expression in (6.50) obviously can not be the spectral density of a well-defined WSS stochastic process: A WSS process having the spectral density in (6.50) has *infinite* variance. The formula in (2.131) assumes that the LTI system is stable so that the output of the LTI system is guaranteed to be WSS when the input is WSS. An ideal integrator is *not* a stable LTI system. When a WSS process is the input to an integrator, the output is, in general, *not* a WSS process. This is very similar to what has been done for $1/f$ noise: Trying to model an inherently nonstationary phenomenon such as $1/f$ noise, or phase noise in free running oscillators, using notions and techniques from the theory of WSS processes is not appropriate.

6.3 Phase Noise: Same at All Nodes

In the phase noise characterization of the oscillator circuits in the previous section, we calculated the noise variance of circuit variables for several oscillator circuits, and found out that the variance waveforms, which were obtained by numerically solving (5.75), had *linear ramp* envelopes. In Section 5.6, we gave an analytical expression for the solution of (5.75), which we repeat below

$$\mathbf{K}(t) = \mathbf{\Phi}(t,0)\mathbf{K}_0\mathbf{\Phi}(t,0)^T + \int_0^t \mathbf{\Phi}(t,\tau)\mathbf{F}(\tau)\mathbf{F}(\tau)^T\mathbf{\Phi}(t,\tau)^T d\tau \qquad (6.51)$$

where $\mathbf{\Phi}(t,\tau)$ is the state transition matrix for the system of linear time-varying homogeneous differential equations

$$\dot{\mathbf{y}} = \mathbf{E}(t)\mathbf{y}. \qquad (6.52)$$

Lemma 6.1 *For an oscillator circuit with a stable steady-state oscillation, one of the eigenvalues of the state transition matrix $\Phi(T,0)$ of the homogeneous system of equations in (6.52) is 1.*

Proof. We assume that (6.10) for the oscillator circuit has a *periodic steady-state* solution $\mathbf{x}_s(t)$, which is a stable limit cycle in the n-dimensional space. Hence,

$$\mathbf{I}(\mathbf{x}_s) + \frac{d}{dt}\mathbf{Q}(\mathbf{x}_s) = 0 \tag{6.53}$$

holds. (6.53) can be rewritten as

$$\mathbf{I}(\mathbf{x}_s) + \mathbf{C}(\mathbf{x}_s)\dot{\mathbf{x}}_s = 0 \tag{6.54}$$

where $\mathbf{C}(\mathbf{x}_s)$ is the Jacobian of $\mathbf{Q}(\mathbf{x}_s)$ as defined by (5.26). Let us take the time derivative of both sides of (6.53) to obtain

$$\mathbf{G}(\mathbf{x}_s)\dot{\mathbf{x}}_s + \frac{d}{dt}[\mathbf{C}(\mathbf{x}_s)\dot{\mathbf{x}}_s] = 0 \tag{6.55}$$

where $\mathbf{G}(\mathbf{x}_s)$ is the Jacobian of $\mathbf{I}(\mathbf{x}_s)$ as defined by (5.19). In this case, t does not appear explicitly as an argument of $\mathbf{G}$, because the circuit is autonomous and it does not have external time-varying excitations. If we expand the derivative in (6.55), we get

$$\mathbf{G}(\mathbf{x}_s)\dot{\mathbf{x}}_s + [\frac{d}{dt}\mathbf{C}(\mathbf{x}_s)]\dot{\mathbf{x}}_s + \mathbf{C}(\mathbf{x}_s)\ddot{\mathbf{x}}_s = 0 \tag{6.56}$$

and hence

$$[\mathbf{G}(\mathbf{x}_s) + \frac{d}{dt}\mathbf{C}(\mathbf{x}_s)]\dot{\mathbf{x}}_s + \mathbf{C}(\mathbf{x}_s)\ddot{\mathbf{x}}_s = 0. \tag{6.57}$$

Then, we substitute $\mathbf{A}(t)$ and $\mathbf{C}(t)$ into (6.57), which were defined by (5.32) and (5.39) in terms of $\mathbf{G}(\mathbf{x}_s)$ and $\mathbf{C}(\mathbf{x}_s)$, to obtain

$$\mathbf{A}(t)\dot{\mathbf{x}}_s + \mathbf{C}(t)\ddot{\mathbf{x}}_s = 0. \tag{6.58}$$

Thus, the time derivative of the periodic steady-state solution, i.e. $\dot{\mathbf{x}}_s$, is a solution of the following *homogeneous* system of linear periodically time-varying differential equations

$$\mathbf{A}(t)\mathbf{y} + \mathbf{C}(t)\frac{d}{dt}\mathbf{y} = 0 \tag{6.59}$$

which represents the LPTV system that is obtained by *linearizing* the nonlinear autonomous oscillator circuit around the periodic steady-state solution. Since $\mathbf{x}_s$ is periodic, $\dot{\mathbf{x}}_s$ is also periodic. Thus, (6.59) has a *periodic steady-state* solution. Now, we can easily show that the homogeneous system (6.52) also has a periodic steady-solution, because it was obtained from (6.59) by eliminating the pure algebraic equations to remove the rank deficiency of $\mathbf{C}(t)$, which was explained in detail in Section 5.4. The state vector for (6.52) is a reduced version of the state vector for (6.59). Hence, the periodic steady-state solution of (6.52) is obtained from $\mathbf{x}_s$ by eliminating some of the variables.

As we saw in Section 2.4.9, the LPTV system (6.52) has a periodic solution if and only if $\mathbf{\Phi}(T,0)$ has an eigenvalue equal to 1, where $\mathbf{\Phi}(t,\tau)$ is the state transition matrix for (6.52), and T is the period of oscillation.□

Furthermore,

Remark 6.2 *The rest of the possibly complex eigenvalues of $\mathbf{\Phi}(T,0)$ should have magnitudes less than 1, if the oscillator is in a stable limit cycle.*

We state the stability condition in Remark 6.2 without a formal proof. Intuitively, it is easy to see the validity of this statement: (6.52) represents an LPTV system that is obtained by *linearizing* the nonlinear autonomous oscillator circuit around the periodic steady-state solution. If $\mathbf{\Phi}(T,0)$ has eigenvalues with magnitudes greater than 1, or more than one eigenvalue with a magnitude that is equal to 1, the homogeneous system (6.52) can have solutions that grow without bound, implying that small perturbations on the periodic steady-state trajectory can cause the oscillator to drift away from the limit cycle, which in turn implies an *unstable* oscillation. For most oscillator circuits, $\mathbf{\Phi}(T,0)$ will have an eigenvalue that is equal to 1, and the rest of the eigenvalues will have magnitudes that are "much" smaller than 1. A second eigenvalue that has a magnitude close to 1 suggests that the oscillator circuit is close to being unstable, which is usually the case for high-Q oscillators [16]. We will calculate the eigenvalues of $\mathbf{\Phi}(T,0)$ for several oscillators later.

Theorem 6.1 *The noise variance of the kth state variable of the circuit is given by*

$$\mathbf{e}_k^T \mathbf{K}(t) \mathbf{e}_k = \dot{x}_s(t)^2 \int_0^t \mathbf{v}_1^T(\tau) \mathbf{F}(\tau) \mathbf{F}(\tau)^T \mathbf{v}_1(\tau) d\tau. \tag{6.60}$$

where

$$\mathbf{e}_k = \begin{bmatrix} 0 & \cdots & 0 & 1 & 0 & \cdots & 0 \end{bmatrix}^T \tag{6.61}$$

with 1 as the kth entry, $\dot{x}_s(t)^2$ is the time derivative of the large signal periodic steady-state waveform for the kth state variable, and $\mathbf{v}_1^T(\tau)$ is periodic vector with period T.

Proof. We first consider the representation of the state transition matrix for the LPTV system (6.52) that was derived in Section 2.4.9 and given by (2.195), which we repeat below

$$\Phi(t,\tau) = \sum_{i=1}^{n} \exp\left(\eta_i(t-\tau)\right)\mathbf{u}_i(t)\mathbf{v}_i^T(\tau) \tag{6.62}$$

where η_i are the Floquet exponents for (6.52), and $\mathbf{u}_i(t)$ and $\mathbf{v}_i(t)$ are vectors that are periodic with T. We know that $\Phi(T,0)$ has an eigenvalue equal to 1, we arbitrarily set[3]

$$\eta_1 = 0. \tag{6.63}$$

With (6.63), we know from Section 2.4.9 that $\mathbf{u}_1(t)$ is a periodic steady-state solution of (6.52). However we also know that $\dot{\mathbf{x}}_s$ is a periodic steady-state solution of (6.52). Thus, we set[4]

$$\mathbf{u}_1(t) = \dot{\mathbf{x}}_s(t). \tag{6.64}$$

Moreover, since $\eta_1 = 0$, $\mathbf{v}_1(t)$ is a periodic steady-state solution of

$$\dot{\mathbf{y}} = -\mathbf{E}(t)^T\mathbf{y} \tag{6.65}$$

which also follows from our discussion in Section 2.4.9.

Next, we use the representation of the state transition matrix in (6.62) for (6.52), to calculate the analytical solution of (5.75) given in (6.51). For oscillator circuits, we set the initial condition to zero

$$\mathbf{K}_0 = \mathbf{K}(0) = \mathbf{0}. \tag{6.66}$$

We will calculate the kth diagonal entry of $\mathbf{K}(t)$, which is the noise variance of the kth state variable that is either an inductor current or a node voltage with a capacitive path to ground. The kth diagonal entry of $\mathbf{K}(t)$ is given by

$$\mathbf{e}_k^T \mathbf{K}(t)\mathbf{e}_k. \tag{6.67}$$

From (6.51) and (6.66)

$$\mathbf{e}_k^T \mathbf{K}(t)\mathbf{e}_k = \int_0^t \mathbf{e}_k^T \Phi(t,\tau)\mathbf{F}(\tau)\mathbf{F}(\tau)^T \Phi(t,\tau)^T \mathbf{e}_k \, d\tau \tag{6.68}$$

from which it follows that

$$\mathbf{e}_k^T \mathbf{K}(t)\mathbf{e}_k = \int_0^t \mathbf{e}_k^T \Phi(t,\tau)\mathbf{F}(\tau)\mathbf{F}(\tau)^T [\mathbf{e}_k^T \Phi(t,\tau)]^T \, d\tau \tag{6.69}$$

We then calculate

$$\mathbf{e}_k^T \mathbf{\Phi}(t,\tau) = \sum_{i=1}^{n} \exp(\eta_i(t-\tau))\mathbf{e}_k^T \mathbf{u}_i(t)\mathbf{v}_i^T(\tau) \qquad (6.70)$$

$$= \sum_{i=1}^{n} \exp(\eta_i(t-\tau))\{\mathbf{u}_i(t)\}_k \mathbf{v}_i^T(\tau) \qquad (6.71)$$

where $\{\mathbf{u}_i(t)\}_k$ denotes the kth entry of $\mathbf{u}_i(t)$. We have $\eta_1 = 0$, and we assume that

$$|\exp(\eta_i)| \ll 1 \quad for \; i=2,\ldots,n \qquad (6.72)$$

since η_i have negative real parts for $i=2,\ldots,n$. (6.72) is satisfied for "most" oscillator circuits. We will later discuss the cases for which (6.72) is not satisfied. When (6.72) is satisfied, the contribution of the terms for $i=2,\ldots,n$ in the summation (6.71) to the integral in (6.69) will be negligible. The value of the integral will be determined by the term for $i=1$ as a quite accurate approximation. Then, (6.71) can be approximated with

$$\mathbf{e}_k^T \mathbf{\Phi}(t,\tau) \approx \exp(\eta_1(t-\tau))\{\mathbf{u}_1(t)\}_k \mathbf{v}_1^T(\tau) \qquad (6.73)$$

$$= \{\mathbf{u}_1(t)\}_k \mathbf{v}_1^T(\tau) \qquad (6.74)$$

to evaluate the integral in (6.69). We substitute (6.74) in (6.69) to obtain

$$\mathbf{e}_k^T \mathbf{K}(t)\mathbf{e}_k = \int_0^t \{\mathbf{u}_1(t)\}_k \mathbf{v}_1^T(\tau)\mathbf{F}(\tau)\mathbf{F}(\tau)^T[\{\mathbf{u}_1(t)\}_k \mathbf{v}_1^T(\tau)]^T d\tau \qquad (6.75)$$

and hence

$$\mathbf{e}_k^T \mathbf{K}(t)\mathbf{e}_k = \int_0^t \dot{x}_s(t)^2 \mathbf{v}_1^T(\tau)\mathbf{F}(\tau)\mathbf{F}(\tau)^T \mathbf{v}_1(\tau) d\tau \qquad (6.76)$$

where we substituted

$$\dot{x}_s(t) = \{\mathbf{u}_1(t)\}_k = \{\dot{\mathbf{x}}_s(t)\}_k \qquad (6.77)$$

using (6.64). We observe that $\mathbf{v}_1^T(\tau)\mathbf{F}(\tau)\mathbf{F}(\tau)^T \mathbf{v}_1(\tau)$ in (6.76) is periodic in τ with period T and satisfies

$$\mathbf{v}_1^T(\tau)\mathbf{F}(\tau)\mathbf{F}(\tau)^T \mathbf{v}_1(\tau) \geq 0 \qquad (6.78)$$

for all $\tau \geq 0$, because $\mathbf{F}(\tau)\mathbf{F}(\tau)^T$ is a positive semidefinite matrix for all $\tau \geq 0$. Since $\dot{x}_s(t)^2$ is independent of the integration variable τ, it can be taken out of the integral, and we obtain

$$\mathbf{e}_k^T \mathbf{K}(t)\mathbf{e}_k = \dot{x}_s(t)^2 \int_0^t \mathbf{v}_1^T(\tau)\mathbf{F}(\tau)\mathbf{F}(\tau)^T \mathbf{v}_1(\tau) d\tau. \qquad (6.79)$$

□

Thus, we obtained an analytical expression for the noise variance of the kth state variable of the circuit. Recall that we can numerically solve (5.75) to calculate the noise variance of circuit variables as a function of time. We did this for the negative resistance oscillator in Section 5.10.4, and obtained the waveform in Figure 5.12. (6.79) tells us that the variance waveform in Figure 5.12 can be approximately constructed by "modulating" the periodic waveform[5] $\dot{x}_s(t)^2$ with the waveform that is given by the integral

$$\int_0^t \mathbf{v}_1^T(\tau)\mathbf{F}(\tau)\mathbf{F}(\tau)^T\mathbf{v}_1(\tau)d\tau \tag{6.80}$$

where the integrand is a nonnegative-valued periodic function of τ. We can also see from (6.80) that the variance waveform will have a *linear ramp* envelope. Moreover, we observe that the peaks in the variance waveform are going to be at the time points where the peaks in $\dot{x}_s(t)^2$ occur, i.e. at the transition or zero-crossing points of the large-signal steady-state periodic oscillation waveform, which are identified as the time points where $\dot{x}_s(t)^2$ takes its largest value. The set Γ we defined with (6.17) contains these time points. We can actually calculate the slope α of the linear ramp envelope as follows:

$$\alpha = \left[\max_{0 \leq t \leq T} (\dot{x}_s(t)^2)\right] \frac{1}{T}\int_0^T \mathbf{v}_1^T(\tau)\mathbf{F}(\tau)\mathbf{F}(\tau)^T\mathbf{v}_1(\tau)d\tau. \tag{6.81}$$

Recall that we used the slope α in Section 6.2.2 to characterize phase noise as a random walk process. The autocorrelation function of phase noise represented as a random walk process was given by (6.31), which we repeat below

$$\mathrm{E}\left[\phi[k]\phi[m]\right] = \beta^2 \alpha \min(k,m)T \quad \text{for} \quad k,m \geq 0 \tag{6.82}$$

where

$$\beta = \frac{2\pi f_c}{S} \tag{6.83}$$

as defined by (6.24). Recall the definition of S as

$$S = \max_{0 \leq t \leq T} \dot{x}_s(t). \tag{6.84}$$

Let us now use (6.81), (6.83) and (6.84) to calculate $\beta^2\alpha$:

$$\beta^2\alpha = (2\pi f_c)^2 \frac{\max_{0 \leq t \leq T}(\dot{x}_s(t)^2)}{[\max_{0 \leq t \leq T}\dot{x}_s(t)]^2}\frac{1}{T}\int_0^T \mathbf{v}_1^T(\tau)\mathbf{F}(\tau)\mathbf{F}(\tau)^T\mathbf{v}_1(\tau)d\tau, \tag{6.85}$$

from which it follows that

$$\beta^2\alpha = (2\pi f_c)^2 \frac{1}{T}\int_0^T \mathbf{v}_1^T(\tau)\mathbf{F}(\tau)\mathbf{F}(\tau)^T\mathbf{v}_1(\tau)d\tau. \tag{6.86}$$

Hence, we conclude with the major result of this section

Corollary 6.1 *$\beta^2 \alpha$ in (6.86) is independent of $\dot{x}_s(t)$. Thus, the autocorrelation of phase noise in (6.82) is independent of $\dot{x}_s(t)$.*

Recall that the phase noise characterization was obtained for a specific circuit variable. The slope for the envelope of the noise variance α was calculated for a specific state variable of the system. As seen in (6.81), α depends on $\dot{x}_s(t)$ which is the derivative of the large-signal periodic steady-solution for this specific state variable as defined by (6.77). However, even though both α and β depend on $\dot{x}_s(t)$, $\beta^2 \alpha$ is independent of it. Thus, the phase noise autocorrelation (6.82) *is* independent of the state variable we choose, making phase noise a property of the whole oscillator circuit instead of only the state variable it was calculated for. This result we arrived at is not obvious for the phase noise definition we presented, and is by no means a trivial observation. Recall that we "experimentally" observed this result for the harmonic oscillator in Section 6.2.3.3.

(6.86) also suggests a way to directly calculate $\beta^2 \alpha$ without first numerically solving (5.75), if we can somehow calculate the periodic vector $\mathbf{v}_1(t)$ for $0 \leq t \leq T$. The periodic vector $\mathbf{v}_1(t)$ can be computed as a periodic steady-state solution of (6.65). It turns out that the phase noise characterization methodology proposed by Franz Kaertner [16] arrives at exactly the same phase noise characterization represented by (6.86) we obtained, even though his definition of phase noise, and his derivation of the phase noise characterization is completely different than ours. The numerical method he proposes for phase noise characterization is based on calculating $\mathbf{v}_1(t)$ and using (6.86). We will discuss Kaertner's work in more detail in the next section.

6.4 Kaertner's Work on Phase Noise

Kaertner [16][6] defines timing jitter/phase noise for a free running autonomous oscillator as a continuous-time stochastic process:

$$\mathbf{X}(t) = \mathbf{x}_s(t + \nu(t)) + \Delta \mathbf{X}(t + \nu(t)) \tag{6.87}$$

where $\mathbf{x}_s(t)$ represents the periodic steady state solution of the oscillator circuit without the noise sources (which is a limit cycle in the n-dimensional space), $\mathbf{X}(t)$ is the trajectory of the oscillator with the noise sources, $\nu(t)$ represents the phase noise/timing jitter as a stochastic *time shift*, and he calls $\Delta \mathbf{X}(t)$ the *amplitude noise* process. Obviously, the decomposition of the noisy trajectory $\mathbf{X}(t)$ as in (6.87) is not unique, and needs further specification. (6.87) can be interpreted as the decomposition of the "difference" between the noisy trajectory $\mathbf{X}(t)$ and the noiseless limit cycle $\mathbf{x}_s(t)$ into a *tangential* motion $\nu(t)$ and a *transversal* motion $\Delta \mathbf{X}(t)$. Kaertner arbitrarily restricts the transversal deviation $\Delta \mathbf{X}(t)$ onto the hyperplane that is spanned by the vectors $\mathbf{u}_2(t), \mathbf{u}_3(t), \ldots, \mathbf{u}_n(t)$ in (6.62).[7] Thus, he obtains

$$\mathbf{v}_1(t)^T \Delta \mathbf{X}(t) = 0 \tag{6.88}$$

using the orthogonality relations given in (2.194). Using orthogonal projection operations which are based on (2.194), he proceeds to derive the following stochastic "integral" equation for the stochastic time shift $\nu(t)$ that represents phase noise:

$$\nu(t) = \int_0^{t+\nu(t)} \mathbf{v}_1^T(\tau) \mathbf{F}(\tau) d\mathbf{W}(\tau). \tag{6.89}$$

Then, he derives a "spectrum" expression for phase noise[8] starting from (6.89). After some "juggling", it can be shown that the phase noise spectrum he derives is exactly equal to (6.50) with $\beta^2 \alpha$ given by (6.86).

Phase noise is a concept with a lot of confusion and controversy around it, because it can be defined in many ways, and it can be characterized in various different ways. We found Kaertner's work on phase noise to be one of the very few that treats the problem using non *ad-hoc* mathematical techniques, and with clear definitions.[9] We believe that Kaertner's treatment of phase noise and the treatment we presented complement each other by producing consistent results with different definitions and derivations.

6.5 Alternative Phase Noise Characterization Algorithm

The phase noise characterization for an oscillator boils down to the calculation of $\beta^2 \alpha$ in (6.86), which requires the computation of the periodic vector $\mathbf{v}_1(t)$ for $0 \leq t \leq T$. Kaertner [16] describes a numerical method for the computation of $\mathbf{v}_1(t)$. Without providing details, we will present the outline of the numerical algorithm for computing $\mathbf{v}_1(t)$, which is very similar to the algorithm proposed by Kaertner:

Algorithm 6.2 (Phase Noise/Timing Jitter Characterization)

1. *Compute the large-signal periodic steady-state solution $\mathbf{x}_s(t)$ for $0 \leq t \leq T$.*

2. *Compute the state-transition matrix $\Phi(T, 0)$ by numerically integrating*

$$\dot{\mathbf{Y}} = \mathbf{E}(t)\mathbf{Y}, \quad \mathbf{Y}(0) = \mathbf{I} \tag{6.90}$$

from 0 to T. Note that

$$\Phi(T, 0) = \mathbf{Y}(T). \tag{6.91}$$

3. Compute $\mathbf{u}_1(0)$ using
$$\mathbf{u}_1(0) = \dot{\mathbf{x}}_s(0). \tag{6.92}$$

   Note that $\mathbf{u}_1(0)$ is an eigenvector of $\Phi(T,0)$ corresponding to the eigenvalue 1.

4. $\mathbf{v}_1(0)$ is an eigenvector of $\Phi(T,0)^T$ corresponding to the eigenvalue 1. To compute $\mathbf{v}_1(0)$, first compute an eigenvector of $\Phi(T,0)^T$ corresponding to the eigenvalue 1, then scale this eigenvector so that
$$\mathbf{v}_1(0)^T \mathbf{u}_1(0) = 1 \tag{6.93}$$

   is satisfied. (6.93) has to be satisfied for the representation of the state transition matrix $\Phi(t,\tau)$ given in (6.62) to be correct, as derived in Section 2.4.9.

5. Compute the periodic vector $\mathbf{v}_1(t)$ for $0 \leq t \leq T$ by numerically solving
$$\dot{\mathbf{y}} = -\mathbf{E}(t)^T \mathbf{y} \tag{6.94}$$

   using $\mathbf{v}_1(0) = \mathbf{v}_1(T)$ as the initial condition. Recall that $\mathbf{v}_1(t)$ is a periodic steady-state solution of (6.94) corresponding to the Floquet exponent that is equal to 0. All of the other Floquet exponents for (6.94) have positive real parts as discussed in Section 2.4.9. Thus, it is not possible to calculate $\mathbf{v}_1(t)$ by numerically integrating (6.94) forward in time, because the numerical errors in computing the solution and the numerical errors in the initial condition $\mathbf{v}_1(0)$ will excite the modes of the solution with Floquet exponents that have positive real parts. However one can integrate (6.94) backwards in time with the "initial" condition $\mathbf{v}_1(T) = \mathbf{v}_1(0)$ to calculate $\mathbf{v}_1(t)$ for $0 \leq t \leq T$ in a numerically stable way.

6. Then, $\beta^2 \alpha$ is calculated using (6.86). Thus, the the discrete-time process ϕ, i.e. phase noise, is fully characterized as a random walk process with the autocorrelation function given in (6.82).

The algorithm we described above is an alternative to the one we have presented in Section 6.2.2. The phase noise characterizations calculated with both of the algorithms should agree, because, in Section 6.3, we have shown that the above algorithm can be derived from the one that was described in Section 6.2.2.[10] The algorithm described in Section 6.2.2 was based on using the general time-domain noise analysis (Chapter 5) technique on an oscillator circuit. Then, using some properties of oscillator circuits, we were able to derive the above algorithm which enables us to calculate the phase noise characterization efficiently without performing a full time-domain noise analysis. This specific phase noise characterization algorithm turns out to be exactly equivalent to the one proposed by Kaertner [16] with a different definition for phase noise, and a different derivation.

We implemented the above algorithm in SPICE. We will not present a detailed description of this implementation here, but we will mention a few important points. In implementing the above algorithm, one can increase the efficiency by saving LU factored matrices that needs to be calculated in Step 2 and reuse them in Step 5. If the large-signal periodic steady-state of the oscillator is calculated using the shooting method[11] in Step 1, then the state transition matrix $\Phi(T, 0)$ of the linear time-varying system obtained by linearizing the nonlinear oscillator circuit around the large-signal periodic steady-state is already available. It can be shown that the *Jacobian* of the nonlinear system of equations that is solved in the shooting method (using Newton's method, to calculate the initial condition that results in the periodic steady-state solution) is equal to $\Phi(T, 0) - \mathbf{I}$ [53][54]. Moreover, one can avoid calculating $\Phi(T, 0)$ explicitly, and use matrix-implicit iterative methods both for the shooting method, and at Step 4 to calculate the eigenvector of $\Phi(T, 0)^T$ that corresponds to the eigenvalue 1 [55]. For high-Q oscillators, the iterative methods can run into problems, because $\Phi(T, 0)$ may have several other eigenvalues which are close to 1. In our implementation in SPICE, we explicitly calculate $\Phi(T, 0)$ and perform a full eigenvalue/eigenvector calculation, which will allow us to investigate the properties of the state-transition matrix for various oscillator circuits. Oscillator circuits are usually small circuits, i.e the dimension of the state vector is not too large. Hence, a full eigenvalue/eigenvector calculation for $\Phi(T, 0)$ is feasible. Even with a full eigenvalue/eigenvector calculation for $\Phi(T, 0)$, the phase noise characterization algorithm discussed above is still very efficient.

We used this implementation to characterize the phase noise of the oscillator circuits we have already characterized with the algorithm in Section 6.2.2. Next, we will present these results and discuss various issues in using this algorithm, and comment on the results we obtain. In the presentation of the examples, we will refer to the phase noise characterization algorithm presented in Section 6.2.2 as *Algorithm I*, and the phase noise characterization algorithm presented in Section 6.5 as *Algorithm II*.

6.5.1 Examples

We first applied this algorithm to the simple negative resistance oscillator we considered in Section 5.10.4. This circuit has only two state variables, i.e. the inductor current and the capacitor voltage. We computed the 2×2 $\Phi(T, 0)$ for this circuit, and performed an eigenvalue/eigenvector calculation. The two

	Ring Oscillator	Relaxation Oscillator
State vector dimension	105	44
Eigenvalues		
1	0.974039	1.134139
2	6.676138e-3	3.941591e-16
3	-5.507e-3 + j 7.128e-4	ϵ
4	-5.507e-3 - j 7.128e-4	ϵ
5	2.475e-4 + j 2.946e-3	ϵ
6	2.475e-4 - j 2.946e-3	ϵ
7	5.393619e-4	ϵ

Table 6.3. Eigenvalues of $\Phi(T,0)$ for the ring and the relaxation oscillators

eigenvalues computed are

$$\mu_1 = 0.996652$$
$$\mu_2 = 0.033287$$

As expected, one of the eigenvalues is very close to 1.[12] The other one is almost two orders of magnitude smaller than 1, justifying our assumption in the derivation of *Algorithm II*. We computed $\beta^2 \alpha$ using *Algorithm II* and obtained the *same* result as the one that was obtained with *Algorithm I* in Section 6.2.2.

We calculated the state-transition matrix $\Phi(T,0)$ for both the ring-oscillator considered in Section 6.2.3.1 and the relaxation oscillator considered in Section 6.2.3.2, and performed a full eigenvalue/eigenvector calculation. The first seven eigenvalues (in order of decreasing magnitude) with magnitudes larger than the machine ϵ are shown in Table 6.3. As expected, one of the eigenvalues for both of the oscillators is close to 1. The second largest eigenvalue for the ring oscillator is more than two orders of magnitude smaller than 1, whereas for the relaxation oscillator it is 15 orders of magnitude smaller, justifying our assumption in using *Algorithm II* for phase noise characterization. The $\beta^2 \alpha$ values calculated using *Algorithm II*, and the ones obtained with *Algorithm I* in Section 6.2.3.1 and Section 6.2.3.2 were the *same*.

	Harmonic Oscillator
State vector dimension	16
Eigenvalues	
1	9.654e-1 + j 1.402e-2
2	9.654e-1 - j 1.402e-2
3	9.985166e-1
4	9.99999999827e-1 + j 1.886e-5
5	9.99999999827e-1 - j 1.886e-5
6	9.99999999982e-1
7	9.99999899752e-1
8	7.068e-1 + j 5.030e-1
9	7.068e-1 - j 5.030e-1
10	4.878e-1
11	3.110e-2

Table 6.4. Eigenvalues of $\Phi(T,0)$ for the harmonic oscillator

We now consider the harmonic oscillator of Section 6.2.3.3. Table 6.4 shows eleven of the eigenvalues for the state-transition matrix $\Phi(T,0)$. The other five eigenvalues were smaller than the machine ϵ. We observe that $\Phi(T,0)$ has 7 eigenvalues with magnitudes that are very close to 1, some of them as complex conjugate pairs. Only one of these eigenvalues is the one that is *theoretically* equal to 1 as we know from the Floquet theory of an oscillator with a stable limit cycle. All the others have magnitudes less than 1. Because of various numerical error sources (mainly the time discretization), the eigenvalues are not calculated precisely. Since this circuit is a high-Q one, $\Phi(T,0)$ has eigenvalues with magnitudes close to 1 other than the one which is supposed to be equal to 1 as obtained theoretically. Because of numerical errors, we can not identify the eigenvalue that is supposed to be equal to 1 theoretically.[13] For the oscillator circuits we have considered up to now, all of the other eigenvalues have magnitudes much smaller than 1, so we were not faced with the problem of identifying the correct eigenvalue. At Step 4 of *Algorithm II*, we need to identify this eigenvalue so that we can choose the corresponding eigenvector of $\Phi(T,0)^T$ as $\mathbf{v}_1(0)$. As seen in Table 6.4, the eigenvalues with magnitudes close to 1 are very "close" to each other, so it is not feasible to identify the correct eigenvalue by calculating $\Phi(T,0)$ more accurately by taking smaller time steps. Instead, we use the following relationship to identify the correct eigenvalue

$$\mathbf{v}_i(0)^T \mathbf{u}_1(0) = \mathbf{v}_i(0)^T \dot{\mathbf{x}}_s(0)$$
$$= \begin{cases} 1 & i = 1 \\ 0 & i \neq 1 \end{cases} \quad (6.95)$$

which follows from (2.148), since $\mathbf{v}_i(0)$ are the eigenvectors of $\Phi(T,0)^T$ and $\mathbf{u}_i(0)$ are the eigenvectors of $\Phi(T,0)$ corresponding to the same eigenvalues. We first calculate $\mathbf{u}_1(0) = \dot{\mathbf{x}}_s(0)$ from the large-signal periodic steady-state solution $\mathbf{x}_s(t)$. Next, we perform a full eigenvalue-eigenvector decomposition of $\Phi(T,0)^T$, and select the eigenvectors with corresponding eigenvalues that are close to 1. Then, we compute the dot products of these eigenvectors with $\mathbf{u}_1(0) = \dot{\mathbf{x}}_s(0)$. Of course, (6.95) will not be satisfied exactly due to numerical errors, but we will still be able to identify the eigenvector that corresponds to the eigenvalue that is theoretically equal to 1. This identification is not feasible by just looking at the eigenvalues. Using the above procedure, we have identified the 3rd eigenvalue in Table 6.4 as the one that is theoretically equal to 1. Then, we used the eigenvector of $\Phi(T,0)^T$ that corresponds to this eigenvalue in *Algorithm II* to characterize the phase noise for the harmonic oscillator, and we obtained the *same* result as the one that was obtained with *Algorithm I* in Section 6.2.3.3. Recall that, in Section 6.3, we derived *Algorithm II* from *Algorithm I* assuming that all the other eigenvalues of $\Phi(T,0)$ other than the one that is exactly equal to 1 have magnitudes "much" less than 1. Clearly, this assumption is not satisfied for the harmonic oscillator, but, still, we obtained the same phase noise characterizations using both *Algorithm I* and *Algorithm II*. We state the following conjecture

Conjecture 6.1 *The derivation of* Algorithm II *from* Algorithm I *is still valid for a high-Q oscillator with a stable limit cycle even when there are other eigenvalues with magnitudes "close" to 1.*

Argument. We do not currently have a proof for this claim, but we will provide some intuition for why we believe this claim is true: Even when the magnitudes of the eigenvalues corresponding to other modes are close to 1, the "directions" of the eigenvectors of $\Phi(T,0)$ corresponding to these eigenvalues will be different enough from the "direction" of the eigenvector, i.e. $\mathbf{u}_1(0) = \dot{\mathbf{x}}_s(0)$, that corresponds to the eigenvalue that is exactly equal to 1. High-Q oscillators are

not the only kind of the oscillator circuits which have other eigenvalues with magnitudes "close" to 1. For instance, a linear time-invariant network, such as the one we used to generate $1/f$ noise in Section 3.4.1, with very large time constants will yield to other eigenvalues of $\Phi(T, 0)$ with magnitudes "close" to 1 with corresponding eigenvectors which have a direction close to $\mathbf{u}_1(0) = \dot{\mathbf{x}}_s(0)$.
□

If there is a second eigenvalue close to 1 with an eigenvector that has a direction close to $\mathbf{u}_1(0) = \dot{\mathbf{x}}_s(0)$, then this suggests that the limit cycle of the oscillator is close to being unstable. This condition can be observed using time-domain noise simulation. The envelope of the noise variance of a state variable for this oscillator will be a *parabola* instead of a linear ramp. We will now present such an example. Let us consider the nonlinear autonomous system that is described by the following state equations:

$$\dot{x} = x - 0.01xy \quad (6.96)$$
$$\dot{y} = -y + 0.02xy \quad (6.97)$$

These state equations describe the Volterra predator-prey model. The limit cycle this system settles into *depends* on the *initial conditions* the system is started at. Hence, this system has an infinite number of limit cycles. Thus, the limit cycle is not stable, and disturbances will cause the nonlinear autonomous system to move from one limit cycle to another. Two limit cycles resulting from the initial conditions

$$x(0) = 20 \quad y(0) = 20$$

and

$$x(0) = 19 \quad y(0) = 19$$

are shown in Figure 6.11. We linearized the system around the limit cycle for the initial conditions

$$x(0) = 20 \quad y(0) = 20$$

and inserted two white noise sources as disturbances to perform a time domain noise simulation. Figure 6.12 shows the noise variance of the state variable x, which has a parabolic envelope as opposed to a linear ramp, which suggests an unstable limit cycle. The noise correlation $\mathbf{E}\left[X_n(t_{ref})X_n(t)\right]$ for $t_{ref} \approx 33.3$ secs is in Figure 6.13, which has a *linear ramp* envelope. This indeed means that the LPTV system has *two* Floquet exponents that are equal to 0. We computed the 2×2 $\Phi(T, 0)$ for this system, and performed an eigenvalue/eigenvector calculation to obtain

$$\mu_1 = 9.99448\mathrm{e}-1 + \mathrm{j}3.137\mathrm{e}-2$$
$$\mu_2 = 9.99448\mathrm{e}-1 - \mathrm{j}3.137\mathrm{e}-2$$

NOISE IN FREE RUNNING OSCILLATORS   199

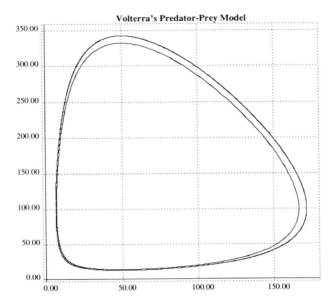

Figure 6.11.   Two limit cycles for the Volterra predator-prey model

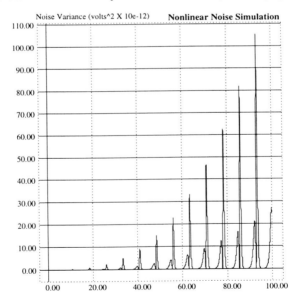

Figure 6.12.   Noise variance for the Volterra predator-prey model

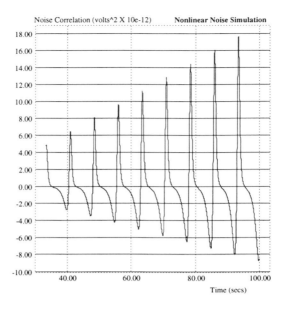

Figure 6.13. Noise correlation for the Volterra predator-prey model

two complex conjugate eigenvalues close to 1! In fact, the system has two eigenvalues at 1, hence unstable. Due to numerical errors, the two eigenvalues calculated are not exactly equal to 1. There is no point in trying to characterize phase noise for this system, because the definition of phase noise for this system does not make sense. Phase noise definition only makes sense for a nonlinear autonomous oscillator circuit that is designed to settle into a stable limit cycle. For an oscillator with a stable limit cycle, phase noise characterizes the errors in the transition times as compared to a noiseless oscillator with transition times separated evenly in time with the period of the oscillation.

6.6 Non-White Noise Sources and Phase Noise

Up to this point in our treatment of phase noise, we considered oscillator circuits with thermal and shot noise sources, i.e. white Gaussian noise sources only. For oscillator circuits with thermal and shot noise, we found out that the phase noise in general can be modeled as a discrete-time random walk process. We obtained this result in general for multiple white noise sources in the oscillator circuit, but we could have obtained it for a single noise source and use the superposition principle to generalize it to the multiple noise source case. We can use the superposition principle for uncorrelated noise sources by summing mean-square noise powers, because we model the oscillator circuit as a *linear system* for noise analysis. For instance, (6.86) can be rewritten as

$$\beta^2 \alpha = (2\pi f_c)^2 \frac{1}{T} \int_0^T \mathbf{v}_1^T(\tau) \mathbf{F}(\tau) \mathbf{F}(\tau)^T \mathbf{v}_1(\tau) d\tau \qquad (6.98)$$

$$= \sum_{i=1}^p (2\pi f_c)^2 \frac{1}{T} \int_0^T [\mathbf{v}_1^T(\tau) \mathbf{f}_i(\tau)]^2 d\tau \qquad (6.99)$$

where p is the number of the noise sources, i.e. the column dimension of $\mathbf{F}$, and $\mathbf{f}_i$ is the ith column of $\mathbf{F}$ which maps the ith noise source to the nodes of the circuit. Hence,

$$(2\pi f_c)^2 \frac{1}{T} \int_0^T [\mathbf{v}_1^T(\tau) \mathbf{f}_i(\tau)]^2 d\tau \qquad (6.100)$$

represents the contribution of the ith noise source to $\beta^2 \alpha$.

For a *white noise* source, phase noise is identified as a random walk process, which can be obtained by sampling a Wiener process, i.e. the integral of a *white noise* process. We can then model the discrete-time phase noise process due to a white noise source as the output of a SISO system which is a cascade of three blocks: an *ideal integrator*, a *gain block* and a *sampler*. The input to the system is a standard stationary white Gaussian noise source. The gain for the gain block is given as the *square root* of (6.100) for a single noise source. If there are multiple white noise sources, this model is still valid when we set the gain to the *square root* of $\beta^2 \alpha$ given in (6.99).

We will now address the following questions:

- Can we model the discrete-time phase noise process due to a *non-white* noise source in the circuit as the output of a SISO system which is a cascade of three blocks: an *ideal integrator*, a *gain block* and a *sampler*. The input to the system is a normalized, i.e scaled, version of the non-white noise source. In other words, is the phase noise model we have described above for a white noise source also valid for a non-white noise source?

- If the answer to the above question is affirmative, how do we calculate the gain of the gain block in the model so that we can characterize the phase noise of an oscillator due to non-white noise sources?[14]

We will answer the above questions for a single noise source that can be obtained by passing a stationary white noise through a low pass filter with a single pole. We will augment the MNA equations for the oscillator circuit so that the low pass filter that is used in the model of the non-white noise source becomes part of the system:

$$I(X) + \frac{d}{dt}Q(X) + bY(t) = 0 \tag{6.101}$$

$$\frac{d}{dt}Y = -\gamma Y + \xi(t) \tag{6.102}$$

where $Y(t)$ represents the low pass filtered white noise source, and the constant vector $\mathbf{b}$ maps this noise source to the nodes of the oscillator circuit. We augment the state vector $\mathbf{X}$ for the oscillator circuit with Y to obtain the new state vector $\mathbf{X}_a$. Now, we can rewrite (6.101) as a system of Ito stochastic differential equations:

$$\mathbf{I}_a(\mathbf{X}_a) + \frac{d}{dt}\mathbf{Q}_a(\mathbf{X}) + \mathbf{B}\xi(t) = 0 \tag{6.103}$$

where

$$\mathbf{X}_a = \begin{bmatrix} \mathbf{X} \\ Y \end{bmatrix} \quad \mathbf{B} = \begin{bmatrix} 0 \\ 1 \end{bmatrix} \tag{6.104}$$

and

$$\mathbf{I}_a(\mathbf{X}_a) = \begin{bmatrix} \mathbf{I}(\mathbf{X}) + \mathbf{b}Y \\ \gamma Y \end{bmatrix}, \quad \mathbf{Q}_a(\mathbf{X}_a) = \begin{bmatrix} \mathbf{Q}(\mathbf{X}) \\ Y \end{bmatrix}. \tag{6.105}$$

If we apply the small noise expansion of Section 5.3 to (6.103), and go through all the steps of Section 5.4 to calculate the augmented versions of $\mathbf{E}$ and $\mathbf{F}$ in (5.75) for the augmented system, we obtain

$$\mathbf{E}_a = \begin{bmatrix} \mathbf{E} & \mathbf{b} \\ 0 & -\gamma \end{bmatrix} \quad \mathbf{F}_a = \begin{bmatrix} 0 \\ 1 \end{bmatrix} \tag{6.106}$$

where $\mathbf{E}$ is for the unaugmented system. Note that $\mathbf{b}$ in (6.106) is not going to be in general equal to the one in (6.105). In fact, even when $\mathbf{b}$ in (6.105) is a constant vector, $\mathbf{b}$ in (6.106) can be time-varying. However we will use the same symbol in order not to complicate the notation.

The augmented $\mathbf{E}_a$ is a periodically time-varying matrix.

Lemma 6.2 *The state transition matrix $\mathbf{\Phi}_a(t,\tau)$ of the homogeneous LPTV system*

$$\dot{\mathbf{y}}_a = \mathbf{E}_a(t)\mathbf{y}_a \tag{6.107}$$

is given by

$$\mathbf{\Phi}_a(t,\tau) = \begin{bmatrix} \mathbf{\Phi}(t,\tau) & \mathbf{a}(t,\tau) \\ 0 & \exp(-\gamma(t-\tau)) \end{bmatrix} \tag{6.108}$$

in terms of the state transition matrix $\mathbf{\Phi}(t,\tau)$ of the LPTV system

$$\dot{\mathbf{y}} = \mathbf{E}(t)\mathbf{y} \tag{6.109}$$

where

$$\mathbf{a}(t,\tau) = \int_\tau^t \mathbf{\Phi}(t,z)\mathbf{b}(z)\exp(-\gamma(z-\tau))dz. \quad (6.110)$$

Proof. (6.108) can be easily verified using the representation of $\mathbf{E}_a$ in (6.106). □

We now proceed as in Section 6.3 to calculate the analytical solution of (5.75) as given in (6.51). For the augmented system, (6.51) takes the form

$$\mathbf{K}_a(t) = \mathbf{\Phi}_a(t,0)\mathbf{K}_{a0}\mathbf{\Phi}_a(t,0)^T + \int_0^t \mathbf{\Phi}_a(t,\tau)\mathbf{F}_a(\tau)\mathbf{F}_a(\tau)^T\mathbf{\Phi}_a(t,\tau)^T d\tau \quad (6.111)$$

We set the initial condition to

$$\mathbf{K}_{a0} = \mathbf{K}_a(0) = \begin{bmatrix} 0 & 0 \\ 0 & \frac{1}{2\gamma} \end{bmatrix} \quad (6.112)$$

In (6.112), we specifically set the initial condition for the variance of the augmented state variable Y to $\frac{1}{2\gamma}$. Recall that the augmented state variable Y represents a low pass filtered white noise. We would like the augmented state Y to represent a WSS process.

Remark 6.3 *If we set the initial condition for the variance of Y (defined by (6.102)) to 0, i.e.* $\mathsf{E}\left[Y(0)^2\right] = 0$, *then it is not a WSS process, and its variance as a function of time is given by*

$$\mathsf{E}\left[Y(t)^2\right] = \frac{1 - \exp(-2\gamma t)}{2\gamma}. \quad (6.113)$$

As seen in (6.113), the steady-state value of the variance for Y is equal to $\frac{1}{2\gamma}$. This is exactly why we set the initial condition to this value so that

$$\mathsf{E}\left[Y(t)^2\right] = \frac{1}{2\gamma} \quad \text{for } t \geq 0 \quad (6.114)$$

is satisfied, and hence, Y is a WSS process that is in steady-state.

Theorem 6.2 *With the choice of the initial condition as in (6.112), the noise variance of the kth state variable of the augmented circuit is given by*

$$\mathbf{e}_k^T \mathbf{K}_a(t)\mathbf{e}_k = \dot{x}_s(t)^2 \left\{ \begin{array}{l} \frac{1}{2\gamma}\left[\int_0^t \mathbf{v}_1^T(z)\mathbf{b}(z)\exp(-\gamma z)dz\right]^2 \\ + \int_0^t \left[\int_\tau^t \mathbf{v}_1^T(z)\mathbf{b}(z)\exp(-\gamma(z-\tau))dz\right]^2 d\tau \end{array} \right\}$$

where
$$\mathbf{e}_k = \begin{bmatrix} 0 & \cdots & 0 & 1 & 0 & \cdots & 0 \end{bmatrix}^T$$
with 1 as the kth entry, $\dot{x}_s(t)^2$ is the time derivative of the large signal periodic steady-state waveform for the kth state variable, and $\mathbf{v}_1^T(z)$ is periodic vector with period T. The kth diagonal entry of $\mathbf{K}_a(t)$ is the noise variance of the kth state variable that is either an inductor current or a node voltage with a capacitive path to ground. We restrict k to be between 1 and n. The $(n+1)$th state variable represents the low pass filtered white noise, and hence the $(n+1)$th diagonal entry of $\mathbf{K}_a(t)$ is given by (6.114).

Proof. We will calculate the first term in (6.111) due to the initial conditions and the second integral term separately. We will refer to the term due to the initial conditions as $\mathbf{K}_{ai}$, and the second term as $\mathbf{K}_{ap}$. We now proceed to calculate $\mathbf{K}_{ai}(t)$:

$$\begin{aligned}
\mathbf{K}_{ai}(t) &= \mathbf{\Phi}_a(t,0)\mathbf{K}_{a0}\mathbf{\Phi}_a(t,0)^T \\
&= \begin{bmatrix} \mathbf{\Phi}(t,0) & \mathbf{a}(t,0) \\ 0 & \exp(-\gamma t) \end{bmatrix} \begin{bmatrix} 0 & 0 \\ 0 & \frac{1}{2\gamma} \end{bmatrix} \begin{bmatrix} \mathbf{\Phi}(t,0) & \mathbf{a}(t,0) \\ 0 & \exp(-\gamma t) \end{bmatrix}^T \\
&= \begin{bmatrix} \frac{\mathbf{a}(t,0)\mathbf{a}(t,0)^T}{2\gamma} & \frac{\exp(-\gamma(t-\tau))\mathbf{a}(t,0)}{2\gamma} \\ \frac{\exp(-\gamma(t-\tau))\mathbf{a}(t,0)^T}{2\gamma} & \frac{\exp(-2\gamma t)}{2\gamma} \end{bmatrix} \quad (6.115)
\end{aligned}$$

where we used (6.108) and (6.112). From (6.115), the kth diagonal entry of $\mathbf{K}_{ai}(t)$ is given by

$$\mathbf{e}_k^T \mathbf{K}_{ai}(t)\mathbf{e}_k = \mathbf{e}_k^T \frac{\mathbf{a}(t,0)\mathbf{a}(t,0)^T}{2\gamma}\mathbf{e}_k \quad (6.116)$$

$$= \frac{a_k(t,0)^2}{2\gamma} \quad (6.117)$$

where $a_k(t,0)$ denotes the kth entry of the $n \times 1$ vector $\mathbf{a}(t,0)$. From (6.110)

$$\mathbf{a}(t,0) = \int_0^t \mathbf{\Phi}(t,z)\mathbf{b}(z)\exp(-\gamma z)\,dz. \quad (6.118)$$

Now, we approximate $\mathbf{\Phi}(t,z)$ in (6.62) with

$$\mathbf{\Phi}(t,z) = \mathbf{u}_1(t)\mathbf{v}_1^T(z) \quad (6.119)$$

as explained in Section 6.3. Then, we substitute (6.119) and (6.118) in (6.117), and after some simple manipulations we obtain

$$\mathbf{e}_k^T \mathbf{K}_{ai}(t)\mathbf{e}_k = \frac{1}{2\gamma}\dot{x}_s(t)^2 \left[\int_0^t \mathbf{v}_1^T(z)\mathbf{b}(z)\exp(-\gamma z)\,dz\right]^2 \quad (6.120)$$

where $\dot{x}_s(t)$ was defined by (6.77) as the kth entry of $\mathbf{u}_1(t)$. Recall that $\mathbf{u}_1(t)$ is the derivative of the large-signal periodic steady-state solution of the oscillator.

We now proceed to calculate $\mathbf{K}_{ap}(t)$, the second term in (6.111):

$$\begin{aligned}
\mathbf{K}_{ap}(t) &= \int_0^t \mathbf{\Phi}_a(t,\tau) \mathbf{F}_a(\tau) \mathbf{F}_a(\tau)^T \mathbf{\Phi}_a(t,\tau)^T d\tau \\
&= \int_0^t \begin{bmatrix} \mathbf{\Phi}(t,\tau) & \mathbf{a}(t,\tau) \\ 0 & e^{-\gamma(t-\tau)} \end{bmatrix} \begin{bmatrix} 0 \\ 1 \end{bmatrix} \begin{bmatrix} 0 \\ 1 \end{bmatrix}^T \begin{bmatrix} \mathbf{\Phi}(t,\tau) & \mathbf{a}(t,\tau) \\ 0 & e^{-\gamma(t-\tau)} \end{bmatrix}^T d\tau \\
&= \int_0^t \begin{bmatrix} \mathbf{a}(t,\tau) \\ \exp(-\gamma(t-\tau)) \end{bmatrix} \begin{bmatrix} \mathbf{a}(t,\tau)^T & \exp(-\gamma(t-\tau)) \end{bmatrix}^T d\tau \quad (6.121)
\end{aligned}$$

where we used (6.108). From (6.121), the kth diagonal entry of $\mathbf{K}_{ap}(t)$ is given by

$$\mathbf{e}_k^T \mathbf{K}_{ap}(t) \mathbf{e}_k = \int_0^t a_k(t,\tau)^2 d\tau \quad (6.122)$$

where $a_k(t,\tau)$ denotes the kth entry of the $n \times 1$ vector $\mathbf{a}(t,\tau)$. Then, we proceed as in the calculation of $\mathbf{K}_{ai}(t)$, and substitute (6.119) and (6.110) in (6.122), and after some simple manipulations we obtain

$$\mathbf{e}_k^T \mathbf{K}_{ap}(t) \mathbf{e}_k = \dot{x}_s(t)^2 \int_0^t \left[\int_\tau^t \mathbf{v}_1^T(z) \mathbf{b}(z) \exp(-\gamma(z-\tau)) dz \right]^2 d\tau \quad (6.123)$$

Now, we calculate the kth diagonal entry of $\mathbf{K}_a(t)$

$$\begin{aligned}
\mathbf{e}_k^T \mathbf{K}_a(t) \mathbf{e}_k &= \mathbf{e}_k^T \mathbf{K}_{ai}(t) \mathbf{e}_k + \mathbf{e}_k^T \mathbf{K}_{ap}(t) \mathbf{e}_k \quad &(6.124) \\
&= \dot{x}_s(t)^2 \quad &(6.125)
\end{aligned}$$

$$\left\{ \frac{1}{2\gamma} \left[\int_0^t \mathbf{v}_1^T(z) \mathbf{b}(z) \exp(-\gamma z) dz \right]^2 + \int_0^t \left[\int_\tau^t \mathbf{v}_1^T(z) \mathbf{b}(z) \exp(-\gamma(z-\tau)) dz \right]^2 d\tau \right\}$$

where $\mathbf{e}_k^T \mathbf{K}_{ai}(t) \mathbf{e}_k$ and $\mathbf{e}_k^T \mathbf{K}_{ap}(t) \mathbf{e}_k$ were substituted from (6.120) and (6.123) respectively. $\square$

We calculated the noise variance of the kth state variable due to a WSS nonwhite noise source that was obtained by low pass filtering a WSS white Gaussian process. Recall that, in Section 6.3, we have done the same calculation for white noise sources and obtained (6.79). We observe that (6.79) and (6.125) are in the same form, i.e. they can be expressed as the product of the periodic $\dot{x}_s(t)^2$ and a modulating function. For multiple white noise sources, the modulating function is given by (6.80) as calculated in Section 6.3. The modulating function for the non-white noise source is given by

$$\frac{1}{2\gamma}\left[\int_0^t \mathbf{v}_1^T(z)\mathbf{b}(z)\exp(-\gamma z)dz\right]^2 + \int_0^t \left[\int_\tau^t \mathbf{v}_1^T(z)\mathbf{b}(z)\exp(-\gamma(z-\tau))dz\right]^2 d\tau \tag{6.126}$$

as seen in (6.125). When the modulating function in (6.80) is sampled at the peak points of $\dot{x}_s(t)^2$, we obtain a linear ramp waveform. This was the reason we were able to model phase noise for white noise sources as the sampled output of an ideal integrator-gain block that is driven by a white noise source. Now, the question is: What do we obtain when we sample the modulating function in (6.126) at the peak points of $\dot{x}_s(t)^2$? Note that $\mathbf{v}_1^T(z)\mathbf{b}(z)$ in (6.126) is a scalar that is a periodic function of its argument z, and recall that $\mathbf{v}_1^T(\tau)\mathbf{F}(\tau)$ in (6.80) is a periodic function of its argument, and it is also a scalar if there is only one noise source in the circuit. Now, we will first evaluate (6.126) by substituting

$$\mathbf{v}_1^T(z)\mathbf{b}(z) = c. \tag{6.127}$$

We will later consider the general case when $\mathbf{v}_1^T(z)\mathbf{b}(z)$ is periodically time-varying. If we substitute (6.127) in (6.126) and evaluate the integrals, we obtain

$$c^2 \left[\frac{-1 + \gamma t + \exp(-\gamma t)}{\gamma^3}\right] \tag{6.128}$$

However (6.128) is *exactly* equal to the variance of a stochastic process that can be obtained as the output of an ideal integrator and a gain block (with gain c) when driven by the same non-white noise source that was in the oscillator:

Lemma 6.3 *The autocorrelation function for the WSS low pass filtered white noise process Y defined by (6.102) is given by*

$$R_Y(\tau) = \frac{\exp(-\gamma|\tau|)}{2\gamma}. \tag{6.129}$$

Theorem 6.3 *Let Y be the input to an integrator, and let Z be the output. Then, the variance of Z as a function of time is given by*

$$\mathrm{E}\left[Z(t)^2\right] = \frac{-1 + \gamma t + \exp(-\gamma t)}{\gamma^3}. \tag{6.130}$$

Proof. The variance of Z is given by [50]

$$\mathrm{E}\left[Z(t)^2\right] = \int_0^t \int_0^t R_Y(t_1 - t_2)dt_1 dt_2. \tag{6.131}$$

from which one can obtain (6.130) by substituting (6.129). □

Thus, we obtain the same expression in (6.128). When $\mathbf{v}_1^T(z)\mathbf{b}(z)$ is a constant c, i.e. when (6.127) is satisfied, we conclude that the phase noise of the oscillator due to a non-white noise source can be modeled as the sampled (at the peak points of $\dot{x}_s(t)^2$) output of a system that is a cascade of an ideal integrator and a gain block with gain c when driven by the same non-white noise source that was in the oscillator. It can be shown that this model is also valid when $\mathbf{v}_1^T(z)\mathbf{b}(z)$ is periodically time-varying. We will discuss the calculation of the gain when $\mathbf{v}_1^T(z)\mathbf{b}(z)$ is periodically time-varying.

We will first give an approximate formula for the square of the gain A without a derivation, and then discuss some special cases to provide some intuition into it. If the periodic $\mathbf{v}_1^T(z)\mathbf{b}(z)$ can be expanded into a Fourier series:

$$\mathbf{v}_1^T(z)\mathbf{b}(z) = \sum_{m=-\infty}^{\infty} c_m \exp(j2\pi m f_c z). \tag{6.132}$$

Then, the formula for the square of the gain is given by:

$$A^2 = (2\pi f_c)^2 \sum_{m=-\infty}^{\infty} |c_m|^2 \frac{\gamma^2}{\gamma^2 + (2\pi m f_c)^2}. \tag{6.133}$$

Let us know consider the case when $\gamma \to \infty$, i.e. the bandwidth of the low pass filter goes to ∞, which means that the noise source approaches a white noise source.[15] Then, (6.133) reduces to

$$A^2 = (2\pi f_c)^2 \sum_{m=-\infty}^{\infty} |c_m|^2 \tag{6.134}$$

$$= (2\pi f_c)^2 \frac{1}{T} \int_0^T [\mathbf{v}_1^T(\tau)\mathbf{b}(\tau)]^2 d\tau \tag{6.135}$$

which is exactly in the same from as (6.100). Recall that (6.100) is the square of the gain for a single white noise excitation. Thus, (6.133) is consistent with our results for white noise sources.

When $\gamma \ll 2\pi f_c$, i.e. when the bandwidth of the noise source is much smaller than the fundamental frequency of the oscillator, (6.133) reduces to

$$A^2 = (2\pi f_c)^2 |c_0|^2. \tag{6.136}$$

We now have a way of characterizing the phase noise of an oscillator due to a non-white noise source. We already have a numerical algorithm to calculate the periodic vector $\mathbf{v}_1^T(z)$. Thus, we can calculate the Fourier series coefficients in (6.132), and use them in (6.133) to calculate the gain of the gain block. Then, phase noise is obtained by sampling the output of a SISO system which is the cascade of an ideal integrator and a gain block driven by the non-white noise source. The sampled phase noise is not a random walk process as it was the case for the white noise source case. Using the above model, we can calculate the autocorrelation function of the sampled phase noise process by first calculating the autocorrelation function of the integrated non-white noise source, then scaling it by the square of the gain given in (6.133) and finally sampling it.

We have seen in Section 3.4.1 that $1/f$ noise sources can be modeled as a summation of Lorentzian spectra, i.e. as the summation of a number of low pass (single pole) filtered white noise sources. So, we can use our results for a low pass filtered white noise source to characterize the phase noise of an oscillator due to a $1/f$ noise source. We accomplish this using the superposition principle.[16] We separately calculate the autocorrelation function of sampled phase noise due to each low pass filtered white noise component in the model of a $1/f$ noise, and then sum these autocorrelation functions to calculate the overall autocorrelation function due to the $1/f$ noise source. For $1/f$ noise sources, noise power will usually be concentrated at frequencies much below the oscillation frequency $2\pi f_c$. Thus, $\gamma \ll 2\pi f_c$ will be satisfied for all of the significant low pass filtered white noise components in the model of a $1/f$ noise source. Then, (6.136) can be used for all of the components. Kaertner in [16] also proposes a phase noise characterization technique for $1/f$ noise sources. After some "digging" into his notation, it can be shown that his results are consistent with what we have concluded above.

Finally, we use superposition to combine the phase noise characterizations we obtained for white and non-white noise sources by separately calculating the autocorrelation functions due to each noise source and then summing them.

6.7 Phase Noise of Phase-Locked Loops

In this chapter, we concentrated on the characterization of phase noise of open-loop, i.e. free running, oscillators. Most often, oscillators are used as the reference oscillator or the voltage/current-controlled oscillator in a phase-locked loop. We would like to be able to characterize the phase noise of oscillators in closed-loop configurations, i.e. when they are placed in phase-locked loops.

If we reexamine the phase noise characterization algorithm we presented in Section 6.2.2 for open-loop oscillators, we can see that it can be easily generalized to characterize the phase noise of closed-loop oscillators. For the closed-loop case, the circuit may not have, and most often will not, have a periodic

NOISE IN FREE RUNNING OSCILLATORS 209

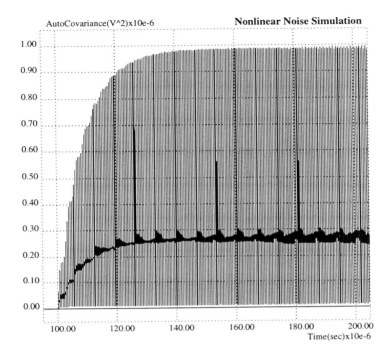

Figure 6.14. Timing capacitor voltage noise variance for the closed-loop multivibrator

large signal steady-state solution. However at steady-state, i.e. when the loop is in lock, the VCO will be in "quasi" periodic steady-state.

We now consider a simple phase-locked loop [1], which uses the relaxation oscillator of Section 6.2.3.2 as the VCO, a Gilbert multiplier type phase detector and a single pole low pass filter as the loop filter. The reference is a noiseless periodic signal onto which the VCO is supposed to lock on. We first simulated this phase-locked loop (at the transistor-level) using transient analysis till the VCO locked onto the reference signal. Hence, we calculated the large signal noiseless steady-state of the circuit. Then, we used the time-domain noise simulation algorithm to calculate the noise variance of the voltage across the timing capacitor in the VCO when the loop is in lock. The variance waveform is in Figure 6.14. In this simulation, the initial condition for the noise variance was chosen to be 0. We observe that the envelope for the noise variance waveform settles to a *steady-state* value. Recall that the noise variance for the open-loop case had a linear ramp envelope that grows without bound. However we are not surprised with this observation, because the phase-locked loop we are considering is not an autonomous system. Even though the noise voltage for the timing capacitor is nonstationary for the open-loop case, it becomes a quasi-cyclostationary process as a result of the negative feedback that comes with the phase-locked loop. Following the phase noise characterization algorithm of Section 6.2.2, we sample the envelope of the noise voltage variance waveform in Figure 6.14, and obtain the variance of the sampled phase noise process as a function of time. The variance of the phase noise process settles to a *time-invariant* steady-state value. Hence, the sampled phase noise is a *WSS* process at steady-state. Recall that the sampled phase noise for the open-loop case is a random walk process, with a variance that increases monotonically with time.[17]

In theory, it is possible to use time-domain noise simulation to characterize the phase noise of oscillators in closed-loop configurations at the transistor-level as we did above for a simple phase-locked loop. However, in practice, there are several difficulties. First of all, even calculating the large-signal steady-state solution of a phase-locked loop can be quite hard. This is due to the fact that a phase-locked loop is a stiff system, i.e. it has widely varying time constants. The time constant of the negative feedback loop is usually much greater than the period of oscillation for the VCO. As a result, one has to simulate for many cycles of the VCO to be able to observe the behavior of the loop. The situation gets worse when the loop has frequency dividers, which are almost always present in phase-locked loops designed for clock generation and frequency synthesis applications. Even if it is possible to calculate the large signal steady-state at the transistor-level, the time-domain noise simulation at the transistor level (with the current numerical algorithms) is not feasible for complicated phase-locked loop circuits with many components. In the next chapter, we will present a *hierarchical behavioral modeling and simulation* approach for phase noise characterization of phase-locked loops. As we will see, this approach not only enables us to characterize the phase noise of signals generated by complicated phase-locked loops in a bottom-up fashion, it also supports the *top-down hierarchical* design of phase-locked loops.

6.8 Summary

In this chapter, we investigated the *phase noise/timing jitter* phenomenon which is of major concern in electronic systems that contain oscillators. We first tried to give some intuition into the problem, and pointed out that both phase noise and timing jitter are related to the noise in the transition or zero-crossing times of an oscillation waveform. We gave a formal definition for phase noise and timing jitter as discrete-time stochastic processes. This definition was given for a general noisy oscillation waveform. Then, we presented a probabilistic characterization algorithm for phase noise based on using the time-domain noise simulation technique of Chapter 5. We used this phase noise characterization algorithm on several practical oscillator circuits, and found out that phase noise of an oscillator can be modeled as a discrete-time random walk process with white noise sources, i.e. thermal and shot noise sources. The phase noise characterization algorithm of Section 6.2.2, i.e. *Algorithm I*, gave us the characterization for a specific state variable of the oscillator circuit. Using results from the Floquet theory of linear periodically time-varying systems, we showed that the characterization that is calculated by *Algorithm I* is in fact independent of the state variable used. Furthermore, we showed that, in general[18], the phase noise of an oscillator with white noise excitations can be modeled as a random walk process, which is consistent with the results we obtained with *Algorithm I* on several practical oscillator circuits. Moreover, the analysis in Section 6.3 resulted in a more specific and efficient algorithm, i.e. *Algorithm II*, to characterize the random walk process that models phase noise. We then reviewed Franz Kaertner's work on phase noise [16], which we believe is one of the few that treats the problem rigorously. We reviewed his definition of phase noise and the characterization algorithm and pointed out that his algorithm is equivalent to *Algorithm II*, even though we derived it using a different definition of phase noise from *Algorithm I*. Our derivation of *Algorithm II* was based on the time-domain noise simulation technique of Chapter 5. Kaertner's definition of phase noise is based on the Floquet theory. Using Floquet theory, we showed the equivalence of his characterization algorithm and ours. We used *Algorithm II* on the same oscillator circuits we characterized with *Algorithm I*, and obtained the same results as expected, since *Algorithm II* was derived from *Algorithm I*. We also discussed various issues in using *Algorithm II*.

We then showed that the phase noise of an oscillator with white noise sources can be modeled as the output of a SISO system that is a cascade of an integrator, a gain block and a sampler driven by a standard white Gaussian noise processes. Then, all we need is to calculate the gain of the gain block to fully characterize phase noise. We pointed out that the gain can be calculated either using *Algorithm I* or *Algorithm II*. We then showed that (again using the techniques of Chapter 5) the same model is valid for a non-white noise source, that is obtained by a single pole low-pass filtering of a white noise source. In this case, the noise source driving the integrator-gain-sampler system is the low-pass filtered white noise source. Hence, the phase noise of an oscillator due to a non-white noise source can not be modeled as a random walk process, but it is modeled as a discrete-time process that can be obtained by sampling a continuous-time process that is the output of an integrator that is driven by the non-white noise source. We then presented the version of *Algorithm II* for a non-white noise source to calculate the gain of the gain block in the phase noise model. We discussed two special cases, when the bandwidth of the non-white noise source is either very large or very small compared with the oscillation frequency. In the limit as the bandwidth of the non-white noise becomes ∞, the characterizations obtained by the two versions of *Algorithm II* (for white and non-white noise sources) were consistent as expected. We also discussed how one can use the above results to characterize the phase noise of an oscillator due to $1/f$ noise sources.

We discussed the phase noise characterization of closed-loop oscillators, i.e. phase-locked loops, and pointed out the applicability of *Algorithm I* to this case, and presented the phase noise characterization of a simple phase-locked loop. We also pointed out some practical difficulties in *transistor-level* phase noise characterization of phase-locked loops, and motivated for a *hierarchical* approach, which is going to be discussed in the next chapter.

Notes

1. The February 1966 issue of the *Proceedings of the IEEE* contains a number of papers on the analysis of noise in oscillators. This issue contains a letter from D.B. Leeson [56] (pages 329-330), which is probably the most cited reference on noise analysis of oscillators. There are also a number of books on phase-locked loops which discuss noise in oscillators, i.e. [50] and [57].

2. The notion of "limit cycle stability" we use throughout this chapter is the *asymptotic orbital stability*. For a precise definition, please see [58].

3. See Section 2.4.9 for the relationship between the Floquet exponents η_i and the eigenvalues of $\mathbf{\Phi}(T,0)$.

4. Note that any scaled version of the periodic steady-state solution (6.52) is also a periodic steady-state solution of (6.52). We have the freedom of choosing the scaling for $\mathbf{u}_1(t)$, and hence setting $\mathbf{u}_1(t) = \dot{\mathbf{x}}_s(t)$, as long as we choose the scaling of $\mathbf{v}_1(t)$ such that (2.194) is satisfied. This is required for the representation (6.62) to be valid.

5. See Figure 6.4 for a plot of $\dot{x}_s(t)$.

6. We will summarize Kaertner's work using the notation and formulation we have introduced, so that the connection between his work on phase noise and ours can be clearly seen.

7. In a former paper [59], Kaertner uses a different restriction to fully specify the decomposition in (6.87). He reexamines this restriction in [16], and modifies it as described above. His motivation for the new definition is that, with the new definition, the decomposition into amplitude and phase noise is independent of the coordinate system used. With his former definition in [59], a change of coordinates also transforms a part of phase noise into amplitude noise and vice versa. The requirement for the invariance of phase noise under coordinate transformations is actually exactly equivalent to the invariance of phase noise for the different state variables of the oscillator circuit we have proved in Section 6.3 for our definition of phase noise.

8. Note that the units of $\nu(t)$ is the units of time, hence it represents the timing jitter process. Thus, if we scale $\nu(t)$ with $2\pi f_c$ we obtain the phase noise process.

9. The reader is referred to [16] for the details of his work, although it might require some "digging" into his notation to clearly see what is going on.

10. Recall that, in this derivation, we have assumed that the eigenvalues of $\Phi(T,0)$ (except the one that is equal to 1 due to the zero Floquet exponent) have magnitudes that is "much" smaller than 1. This assumption is not justified for high-Q circuits. We will see an example for this later.

11. Computing the large-signal periodic steady-state for autonomous circuits is still an active research area, which is especially difficult for high-Q oscillator circuits.

12. Since we calculate the state transition matrix $\Phi(T,0)$ by numerical integration which discretizes time, this eigenvalue is not equal to 1. If we use smaller time steps to calculate $\Phi(T,0)$, then the eigenvalue calculated becomes closer to 1. However the time discretization is not the only numerical error source. Numerical errors also arise due to errors in calculating the oscillation period T and in the numerical calculation of the eigenvalues.

13. The issue of identifying the correct eigenvalue was not addressed by Kaertner in [16].

14. Recall that we presented two algorithms, referred to as *Algorithm I* and *Algorithm II* to calculate the gain, i.e. the square root of $\beta^2 \alpha$, for the white noise case.

15. Actually, to take this limit one has to scale the noise source Y (as defined by (6.102)) by γ so that it becomes a white noise source with unity spectral density in the limit $\gamma \to \infty$. One also has to truncate the Fourier series in (6.132) so that this limit makes sense.

16. We can use the superposition principle, because we model the oscillator as a linear system for noise sources. However the application of superposition to stochastic process excitations is not the same as it is for deterministic excitations. For stochastic process excitations, we use superposition with mean square quantities such as spectral density or autocorrelation functions. The stochastic process excitations need to be *independent* to use superposition.

17. Note that these results were obtained for a noiseless reference signal for the phase-locked loop. If the reference signal is generated by a real oscillator, instead of being an ideal noiseless periodic waveform, the closed-loop VCO phase noise will not be a WSS process. Even though the negative feedback affects the noise generated in the VCO in such a way so that the resulting closed loop phase noise is WSS, it does not have the same affect on the noise generated in the reference oscillator. If the reference oscillator itself is an open-loop circuit with a nonstationary phase noise, this will cause the phase noise of the closed-loop VCO to be nonstationary.

18. For this derivation, we assumed that the Floquet exponents (except the one that is exactly equal to 0) associated with the LPTV system obtained by linearizing the oscillator circuit around the periodic steady-state have magnitudes much less than 1.

7 BEHAVIORAL MODELING AND SIMULATION OF PHASE-LOCKED LOOPS

Phase-locked loops (PLLs) are widely used as clock generators for microprocessors, for the frequency synthesis of the LO (local oscillator) signal in transceivers, etc. Almost every RF transceiver used in wireless communications contains at least one frequency synthesis PLL. PLL circuits can be quite large and complex. They may contain both analog and digital components. A PLL is basically a nonlinear feedback system. PLLs have a unique property that makes their analysis and simulation, even just for their deterministic behavior, quite difficult: A PLL is a *stiff* system with widely-varying time constants.

One of the major concerns in the design of PLLs for clock generation applications is the *timing jitter* of the clock signal produced by the PLL. Accurate and efficient characterization of the timing jitter of the signal produced by the PLL is crucial. For PLLs that are used as frequency synthesizers in transceivers, the spectral purity of the oscillation signal is extremely important. Ideally, one would like the oscillation signal to be at a *constant* frequency with equally spaced transitions. However, due to noise and systematic (i.e. non random) nonidealities in the system, the signal frequency fluctuates and the transitions of the oscillation are not equally spaced in time. Accurate and efficient characterization of the spectral properties of the oscillation signal in the presence of noise and other systematic nonidealities is crucial. We would like to be able to get estimates for the timing jitter/phase noise of the signal as well as the distortion due to systematic nonidealities at the concept phase of our PLL design, so that we can make design decisions and choose and/or develop a PLL architecture that will meet the specifications that are given to us. We would like to be able to accomplish this without having to design all of the PLL components in detail. Once this top-down design process is completed, we would like to be able to simulate the PLL before we fabricate it, and verify that it indeed meets the specifications we have designed it for. Again, at this stage of the design, verification of the timing jitter/phase noise and the distortion performance is crucial.

Our above discussion dictates the following requirements on any modeling and simulation technique for the design of PLLs:

- The model that is to be used to simulate the whole PLL circuit has to be *abstract* enough so that we will be able to see the results in a reasonable amount of time, considering that the simulation of the PLL will be the part of an optimization (manual or automated) loop.

- The model for the simulation of the PLL is generic enough so that we can simulate it even when we do not have the detailed designs of its building blocks.

- The model for the PLL has to capture nonideal effects so that we can use it to estimate the distortion and phase noise/timing jitter of the output.

At first sight, these requirements could seem to be contradictory. We are required to use *abstract* and *generic* models for *efficiency* and *generality*. On the other hand, we are required to capture seemingly low-level effects such as distortion causing nonidealities and the timing jitter/phase noise behavior. In this chapter, we will present models and simulation techniques for PLLs which satisfy the above requirements, which we will refer to as *behavioral* modeling and simulation techniques for PLLs. The behavioral models we will be describing for the timing jitter/phase noise behavior of PLLs will rely on the techniques and results of Chapter 6, where we already discussed the characterization and modeling of timing jitter/phase noise for open-loop oscillators.

The objective of *behavioral modeling*, in general, is to represent circuit functions with *abstract mathematical models* that are independent of implementation details. While there is a great deal of activity to define and develop a standard analog hardware behavioral description language, in our opinion, there is still *little* work being done to develop behavioral models and specialized simulation techniques for analog and mixed-signal circuits. The top-down design process implies a well-defined behavioral description of the analog function. The behavioral characterization of analog circuits is quite different from the digital one; the analog characterization is composed of *not only* the function that the circuit is to perform, but also the *second order non-idealities* intrinsic to analog operation. In fact, errors in the design often stem from the non-ideal behavior of the analog section, not from the selection of the "wrong" functionality. To shorten the design cycle, it is essential that design problems be discovered as early as possible. For this reason, behavioral simulation is an essential component of any methodology for the design of analog and mixed-signal systems. This simulation can help in selecting the correct architecture to implement the analog function with constraints on the amount of non-idealities that is allowable given a set of specifications at the system level. The reader is referred to [60, 4] for a detailed treatment of behavioral modeling and simulation for analog and mixed-signal circuits, and its use within the framework of a top-down constraint-driven design methodology for analog and mixed-signal systems.

In Chapter 4, Chapter 5 and Chapter 6, we discussed the analysis and simulation of noise in electronic circuits and systems that can be described by a system of algebraic and stochastic differential equations in MNA form. The dynamics of any electronic circuit or system, which is modeled as an interconnection of basic network elements such as resistors, capacitors, inductors, controlled sources, independent voltage and current sources, etc., can be described with a system of algebraic and stochastic differential equations in MNA form. Recall that the models of basic semiconductor devices (diodes, transistors, etc.) are expressed as an interconnection of basic network elements. We will refer to the model of an electronic circuit that is described as an interconnection of basic semiconductor devices and other network elements as a *transistor-level* model. In the former chapters, we were mainly dealing with electronic circuits that were described with a transistor-level model. The noise sources in the system are then the shot, thermal and $1/f$ noise sources associated with the semiconductor devices and other dissipative components. On the other hand, the negative resistance oscillator we discussed in Section 5.10.4 was not described by a transistor-level model. The nonlinear resistor in this circuit that has a negative resistance region was described by a *macromodel*, which could have been realized using some interconnection of semiconductor devices and other components. Thus, electronic circuits which contain macromodels as well as semiconductor devices can be described with a system of algebraic and stochastic differential equations in MNA form. Moreover, one can "emulate" the model of any system (which need not be electronic) that is described as a system of stochastic differential equations by using the basic network elements, e.g. nonlinear and linear controlled sources. We already saw an example for this at the end of Section 6.5, the simple Volterra model of a predator-prey system.

Hence, we can perform noise analysis of all kinds of electronic circuits (and other kinds of systems with stochastic excitations that satisfy our assumptions), which are modeled as an interconnection of basic network elements with noise sources that can be modeled in terms of white noise sources. However we saw in Section 6.7 that it may not be feasible or practical to perform a noise analysis for an electronic circuit fully described at the transistor-level. Phase-locked loops (PLLs) used for clock generation and as frequency synthesizers in RF transceivers are such circuits. One idea in trying to overcome the difficulties in analyzing and simulating PLLs was to use *macromodels* for some of the components instead of their full transistor-level description. Traditionally, *macromodeling* for analog circuits is used to mean creating a model of a circuit block from the implementation details (transistor-level circuit configuration) using pre-defined circuit primitives, i.e. basic network components such as linear/nonlinear controlled sources, resistors, capacitors, switches, etc. Actually, all the semiconductor device models available in circuit simulators are implicitly composed of the same circuit primitives, which makes them conceptually equivalent to a macromodel. The transistor-level representation of a circuit block can be thought to be a very detailed "macromodel". In a macromodel, one tries to capture *much more* "functionality" than there is in a single semiconductor device, with *far less* implementation details. This suggests a definite trade-off between accuracy and the complexity (number of components) of the macromodel. Macromodels in circuit simulators are used to reduce the simulation time (because of reduced number of nodes and complexity), and to simulate circuits with subblocks without implementation details.

Macromodel creation (for a specific circuit block described at the transistor level) for nonlinear circuits is done by "iterating" over the below two steps:

- First, a parameterized model (an interconnection of circuit primitives) which captures the "functionality" of the circuit block being modeled is created. There is no systematic way to create these models and this step heavily depends on the experience of the designer.

- Then, the parameters of the created model is optimized (manually or automatically) to meet some specifications.

When macromodels are used for some of the components of a PLL (e.g. phase detectors and frequency dividers), the time it takes to simulate the acquisition of a PLL can come down to *days* from *weeks*, which is still not acceptable if we are going to use the results of these simulations for design decisions. Even when macromodels are used for all of the components of a PLL, simulating just the acquisition behavior can take *hours*. Hence, we need to work with even more *abstract* models of a PLL to be able to simulate it in a reasonable time.

Trying to simulate a PLL with all of its components described at the transistor-level is one extreme. The other extreme is to use an LTI model to describe the PLL circuit, even though it is an inherently nonlinear system. The LTI models, usually referred to as *phase domain* models, have indeed been very useful in understanding the operation of PLLs and investigating various trade-offs in their design. LTI analysis of PLLs is very efficient and it is very useful in the concept phase of a design. However the usefulness of LTI phase domain models are limited by the fact that they are linear and can not capture some nonidealities in the system that arise from nonlinearities and the interaction of nonlinearities with noise.

In this chapter, we present a behavioral specification and simulation methodology for PLLs. *Mathematical abstraction* is the crucial concept behind this methodology. This approach enables one to concentrate on one aspect of the problem, e.g. noise performance. Thus, one can deal with complexity in an efficient and effective way.

7.1 PLLs for Clock Generators and Frequency Synthesizers

In this chapter, we will be concentrating on a particular phase-locked loop architecture that is widely used in clock generation and frequency synthesis applications. Even though we will be describing the behavioral models and the simulation techniques within the context of this architecture, the modeling and the simulation methodology are general, and hence they can be used for other PLL architectures.

The PLL architecture is shown in Figure 7.1. It consists of a voltage-controlled oscillator (VCO), three frequency dividers (FDs), a reference oscillator (RO), a phase-frequency detector (PFD), a charge pump (CP) and a loop filter (LF). Given a reference oscillator (which is usually an off-chip low phase noise crystal oscillator) with a fixed oscillation frequency f_{ref}, when the PLL is in lock, the frequency of oscillation at the output of the PLL is given by

$$f_o = \frac{N}{LM} f_{ref}. \qquad (7.1)$$

By choosing the FD divide ratios appropriately, one can use this PLL to generate oscillation signals at a number of frequencies in a frequency range that is supported by the VCO. In video driver systems, on-chip PLLs with this architecture are used to generate clocks at a number of discrete frequencies. In RF transceivers, PLLs with this architecture are used to generate the local

220 NOISE IN NONLINEAR ELECTRONIC CIRCUITS

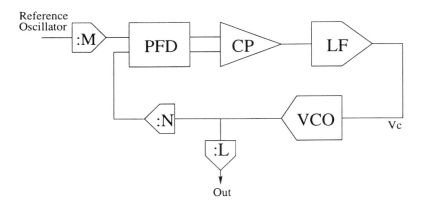

Figure 7.1. Charge-pump phase-locked loop

oscillator (LO) signal which has to have a programmable frequency so that the receiver/transmitter can be tuned to the particular channel of interest.

We will not go into the details of the operation principles of this PLL architecture here, which is covered in many textbooks on PLLs.[1]. It is basically a nonlinear feedback system. It is also a mixed-signal system: PFD is a digital component which is sensitive to only the low-to-high or high-to-low transitions at its inputs. It has two digital outputs which control two switches in the CP. CP and the LF "convert" the digital error information at the output of the PFD into an analog signal that controls the frequency of the VCO. The divide ratio of the FD inside the feedback loop (i.e. the FD with divide ratio N) can be quite large (e.g. 5000). This is one of the main reasons why it is very difficult to simulate this PLL architecture. For a single comparison of the signals at the PFD input, the VCO has to go through *many* cycles of oscillation. When we describe the behavioral models for the components of this PLL, we will see how one can eliminate the high frequency "node" between the VCO and the divider with very little loss of accuracy but with great gain in the efficiency of the simulation.

Now, we would like to discuss a *phase domain* LTI model for this PLL, which is very commonly used to analyze its behavior when the loop is in lock. Our main purpose in discussing the LTI phase domain noise model is to gain a qualitative understanding about the operation of the loop, which will very useful in creating abstractions of the components for behavioral simulation. One of the key requirements for *developing* and *using* behavioral models in the design of a complex mixed-signal system is to have a good understanding of the operation of the system. In the end, what we put into the behavioral models

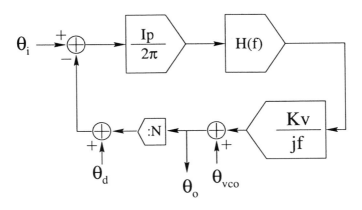

Figure 7.2. LTI phase domain model for the charge-pump PLL

will be our experience and knowledge about the operation of the circuit. The phase domain LTI model for the PLL architecture in Figure 7.1 is shown in Figure 7.2. To simplify the discussion, we omitted the FDs which are outside of the feedback loop. Ideally, when the loop is in lock, without any nonidealities in the system, the frequency of the VCO will be given by

$$f_o = N f_{ref}. \tag{7.2}$$

Since the PFD is a digital component, and is only sensitive to the transitions of the oscillation signals, the particular shape of the waveforms generated by the VCO or the reference oscillator are irrelevant. Thus, the VCO and the reference oscillator are represented with their phases only. The phase of the VCO with a fixed frequency f_o is given by

$$\Phi_o(t) = 2\pi f_o t \tag{7.3}$$

which is a linear function of time. The VCO makes a transition at its output every time the phase crosses an integer multiple of π. The phase variable θ_o in the LTI phase domain model of Figure 7.2 represent deviations from the one in (7.3) due to the noise sources in the system. θ_i in Figure 7.2 represents the phase noise process for the reference oscillator, and θ_{vco} represents the phase noise of the VCO. The phase noise contribution of the FD is represented by θ_d. We assume that the relationship between the frequency of the VCO and the control voltage is given by

$$f_{vco} = f_o + K_v v_c \tag{7.4}$$

where f_o is given by (7.2) and v_c is the control voltage. The PFD is modeled as a gain block with gain $\frac{1}{2\pi}$.[2] CP is modeled as a gain block with gain I_p. The LF is modeled as a SISO LTI system with transfer function $H(f)$. The variable at the output of the VCO represents its phase, and the variable at its input is the control voltage. Phase of the VCO is the integral of its frequency, and its frequency is proportional to the control voltage as given by (7.4). Hence, the VCO is modeled as an integrator with transfer function

$$\frac{K_v}{jf}.$$

The FD is simply modeled as a gain block with gain $\frac{1}{N}$, since its input and output are phase variables. The phase domain LTI model of Figure 7.2 can be considered as a small-signal model obtained by linearization around a steady-state where the reference signal is at frequency f_{ref}, the VCO is at frequency f_o and the control voltage v_c is 0. If $\theta_i = 0$, $\theta_d = 0$ and $\theta_{vco} = 0$, then

$$\theta_o = 0 \qquad v_c = 0$$

is satisfied at steady-state.

We will now calculate the transfer functions from θ_i, θ_d and θ_{vco} to the output θ_o. Since, the system is LTI, we can use the superposition principle. Let us first define

$$G(f) = \frac{1}{2\pi} I_p H(f) \frac{K_v}{jf}. \tag{7.5}$$

To calculate the transfer function $T_i(f)$ from θ_i to θ_o, we set $\theta_d = 0$ and $\theta_{vco} = 0$, and write the following frequency-domain relationship from the model in Figure 7.2:

$$\left[-\frac{\theta_o(f)}{N} + \theta_i(f) \right] G(f) = \theta_o(f). \tag{7.6}$$

From (7.6), it follows that

$$T_i(f) = \frac{\theta_o(f)}{\theta_i(f)} = \frac{G(f)}{1 + \frac{G(f)}{N}}. \tag{7.7}$$

To calculate the transfer function $T_{vco}(f)$ from θ_{vco} to θ_o, we set $\theta_d = 0$ and $\theta_i = 0$, and write the following frequency-domain relationship from the model in Figure 7.2:

$$-\frac{\theta_o(f)}{N} G(f) + \theta_{vco} = \theta_o(f). \tag{7.8}$$

From (7.8), it follows that

$$T_{vco}(f) = \frac{\theta_o(f)}{\theta_{vco}(f)} = \frac{1}{1 + \frac{G(f)}{N}}. \tag{7.9}$$

Similarly, the transfer function $T_d(f)$ from θ_d to θ_o is calculated as

$$T_{vco}(f) = \frac{\theta_o(f)}{\theta_{vcp}(f)} = \frac{-G(f)}{1+\frac{G(f)}{N}}. \qquad (7.10)$$

First, we observe that the transfer functions $T_d(f)$ and $T_i(f)$ are equal.[3] Hence, the phase noise of the reference oscillator and the phase noise contributed by FD have exactly the same effect on the phase noise of the output. We have not shown it in the model of Figure 7.2, but the PFD also contributes to the phase noise of the output. It can be shown that the effect of the phase noise contributed by the PFD has also the same effect on the phase noise of the output as the reference oscillator or the FD. The transfer function from the phase noise contributed by the PFD to the output θ_o is the same as the transfer function from θ_i to θ_o.

The loop filter transfer function $H(f)$ always has a low pass characteristics. Hence, the transfer function $G(f)$ given by (7.5) also has a low pass characteristics, i.e.

$$|G(f)| \to 0 \quad \text{as} \quad f \to \infty.$$

Moreover,

$$|G(f)| \to \infty \quad \text{as} \quad f \to 0$$

i.e. it has infinite gain at $f = 0$. Then, we have

$$|T_i(f)| \to 0 \quad \text{as} \quad f \to \infty$$
$$|T_d(f)| \to 0 \quad \text{as} \quad f \to \infty$$
$$|T_{vco}(f)| \to 1 \quad \text{as} \quad f \to \infty$$

and

$$|T_i(f)| \to N \quad \text{as} \quad f \to 0$$
$$|T_d(f)| \to N \quad \text{as} \quad f \to 0$$
$$|T_{vco}(f)| \to 0 \quad \text{as} \quad f \to 0$$

Hence, we conclude that

- The low frequency phase fluctuations in the reference oscillator signal or the ones contributed by the FD or the PFD are not modified by the feedback loop and appear as phase fluctuations at the output.[4] The high frequency phase fluctuations in the reference oscillator signal or the ones contributed by the FD or the PFD are *rejected* by the feedback loop and do not appear as phase fluctuations at the output.

- The low frequency phase fluctuations contributed by the VCO are *rejected* by the feedback loop, but the high frequency phase fluctuations contributed by the VCO are not modified by the feedback loop and appear as phase fluctuations at the output.

We now have some qualitative understanding on how the PLL reacts to phase fluctuations that are contributed from various sources, which will be extremely useful when we are trying to develop behavioral models, i.e. abstractions, of the PLL components that capture the phase noise behavior. We presented this simple LTI analysis to illustrate the behavioral model development process, i.e how one uses qualitative understanding, or designer experience, about the operation of the system in developing the behavioral models.

7.2 Behavioral Models of PLL Components

We now have a qualitative understanding of the operation of the PLL. We have also extensively investigated the phase noise/timing jitter of open-loop oscillators in Chapter 6. We will now present a description of the behavioral models of the PLL components that capture the phase noise/timing jitter behavior.

7.2.1 *Reference oscillator and the VCO*

The phase detector in the PLL architecture we are considering is a digital one. It is sensitive to only the transitions of the oscillation signals at its input. The particular shapes of the oscillation signals generated by the reference oscillator and the VCO are irrelevant and have very little effect on the operation of the loop. Hence, we model all the oscillation signals generated by the reference oscillator and the VCO as digital signals with high and low states, and transitions between these states, as proposed by Liu in [61]. Let f be the *instantaneous frequency* of the oscillator. The time derivative of the *phase* θ of the oscillation signal is equal to the instantaneous angular frequency. Hence,

$$\dot{\theta}(t) = 2\pi f(t). \tag{7.11}$$

Since the frequency of the oscillator is a positive quantity, the phase θ is a monotonically increasing function of time t. Every time the phase θ crosses an integer multiple of π, the oscillation signal makes a transition, i.e.

$$\theta(t_k) = k\pi \tag{7.12}$$

where t_k is the time for the kth transition of the oscillation signal. At even multiples of π, it makes a low-to-high transition, and at odd multiples of π, it makes a high-to-low transition, or vice versa. For the reference oscillator, the frequency $f(t) = f_{ref}$ does not change with time. Then, the transition times, identified as the time points where the phase θ crosses integer multiples of π, i.e.

$$t_k = \frac{k}{2f_{ref}} \qquad (7.13)$$

will be evenly separated with a separation that is equal to the half of the period of oscillation. Obviously, this model does not capture the *phase noise/timing jitter* behavior of the reference oscillator. To capture the phase noise/timing jitter of the oscillator, we modify (7.12) as follows:

$$\theta(t_k) = k\pi + \phi[k] \qquad (7.14)$$

where ϕ is a discrete-time stochastic process that represents the *phase noise* of the oscillator. In Chapter 6, we have defined and characterized phase noise for open-loop oscillators as discrete-time stochastic processes. We also presented algorithms to calculate the autocorrelation function of the discrete-time phase noise process ϕ. Note that the characterization of phase noise for an oscillator is done for a specific oscillation frequency. The reference oscillator is at a fixed frequency, but the VCO can oscillate at a number of frequencies determined by the divide ratio N of the FD. When the loop is in lock, the VCO oscillation frequency is given by[5]

$$f_o = Nf_{ref}. \qquad (7.15)$$

However, of course, this is true if there are no systematic error sources and there is no noise in the PLL. Typically, the VCO frequency will exhibit fluctuations around the frequency given in (7.15). The VCO frequency is controlled by the signal that is produced by the LF. We model this relationship with a memoryless[6] nonlinear map g

$$f_o = g(v_c). \qquad (7.16)$$

We characterize the VCO phase noise for the single oscillation frequency in (7.15) and use this characterization to model the VCO phase noise when the PLL is in lock.

In summary, we model the reference oscillator with

$$\begin{aligned} \dot{\theta}_{ref}(t) &= 2\pi f_{ref} \\ \theta_{ref}(t_k^{(ref)}) &= k\pi + \phi_{ref}[k] \end{aligned} \qquad (7.17)$$

and the VCO with

$$\begin{aligned} \dot{\theta}_{vco}(t) &= 2\pi g(v_c) \\ \theta_{vco}(t_k^{(vco)}) &= k\pi + \phi_{vco}[k] \end{aligned} \qquad (7.18)$$

where $k = \{0, 1, 2, 3, \ldots\}$. $t_k^{(ref)}$ denotes the kth transition time for the reference oscillator, $t_k^{(vco)}$ denotes the kth transition time for the VCO. v_c is the control voltage that sets the instantaneous oscillation frequency for the VCO. ϕ_{ref} and ϕ_{vco} are discrete-time stochastic processes that represent the phase noise of the reference oscillator and the VCO respectively. The probabilistic characterizations of ϕ_{ref} and ϕ_{vco} are obtained at the frequencies f_{ref} and $f_o = N f_{ref}$ for the reference oscillator and the VCO respectively. These characterizations can be obtained using the techniques of Chapter 6 or using the specification sheets for off-the-shelf components.

In describing the above models for the VCO and the reference oscillator, we did not make any assumptions on the implementation details of the oscillators. Any type of oscillator, e.g. ring-oscillators, relaxation oscillators or LC/resonant oscillators, can be represented using the above oscillator model that captures the phase noise behavior. We made a distinction between the VCO and the reference oscillator, because the reference oscillator has a fixed oscillation frequency, on the other hand, the VCO frequency is controlled by the LF output.

7.2.2 Frequency dividers

FDs are basically digital counters. An FD with divide ratio N transfers every Nth transition at its input to its output, hence it effectively creates an oscillation signal at its output that is at a frequency that is equal to the input signal frequency divided by N.

The input to the FD inside the feedback loop in Figure 7.1 is the VCO output that is modeled by (7.18). Then, the oscillation signal at the output of the FD is modeled by

$$\begin{aligned} \dot{\theta}_{fd}(t) &= \frac{2\pi g(v_c)}{N} \\ \theta_{fd}(t_k^{(fd)}) &= k\pi + \frac{\phi_{vco}[k]}{N} \end{aligned} \quad (7.19)$$

where $k = \{0, 1, 2, 3, \ldots\}$. $t_k^{(fd)}$ denotes the kth transition time for the oscillation signal at the output of the divider. The model in (7.19) is for a noiseless FD that does not contribute any phase noise/timing jitter. We will discuss the modeling of the phase noise contribution of the FD shortly, but we would like to first discuss how a noiseless FD "transfers" the phase noise of the signal at its input to its output. An ideal noiseless FD simply passes every Nth transition at its input to its output. Hence, the *timing jitter* of the transitions at the input are transferred to the output without change. Recall that the timing jitter process J_{vco} for the VCO signal in terms of the phase noise process ϕ_{vco}

is given by
$$J_{vco}[k] = \frac{\phi_{vco}[k]}{2\pi f_o} \qquad (7.20)$$

where f_o is the frequency of the signal at the output of the VCO. Then, the signal at the output of the noiseless FD will have the same timing jitter for the transitions, but the phase noise at the output of the FD will be given by

$$\frac{2\pi f_o}{N} J_{vco}[k] = \frac{\phi_{vco}[k]}{N}. \qquad (7.21)$$

Hence, the phase noise at the output of the FD is smaller than the phase noise at its input by a factor of N, but the timing jitter is the same. On the other hand, the frequency of the signal at the output of the FD is also smaller than the frequency of the signal at its input by a factor of N. Phase noise is a quantity that should be considered relative to the frequency of the signal. In order to compare the phase noise of two oscillation signals, they need to be at the same frequency. Timing jitter can be considered, in a sense, as the "frequency-normalized" version of phase noise. Hence, one can compare the timing jitter of two oscillation signals at different frequencies. It is misleading to compare the phase noise characterizations of the signals at the input and the output of the FD, since it suggests that the FD "improves" the phase noise behavior of the signal. It just seems to be improved, because the signal is just at a lower frequency, but with the same timing jitter around the transitions.

The FD also contributes to the timing jitter/phase noise of the signal at its output. To model this contribution, we modify (7.19) as follows:

$$\begin{aligned} \dot{\theta}_{fd}(t) &= \frac{2\pi g(v_c)}{N} \\ \theta_{fd}(t_k^{(fd)}) &= k\pi + \frac{\phi_{vco}[k]}{N} + \frac{2\pi f_o}{N} J_{fd}[k] \end{aligned} \qquad (7.22)$$

where $k = \{0, 1, 2, 3, \ldots\}$. $t_k^{(fd)}$ denotes the kth transition time for the oscillation signal at the output of the divider. J_{fd} is a discrete-time stochastic process that represents the timing jitter contribution of the FD.

7.2.2.1 Characterization of timing jitter for frequency dividers

Given the implementation of an FD, one can use the time domain noise simulation of Chapter 5 to characterize the timing jitter process J_{fd}. This is done by driving the FD with the large signal output of the VCO to calculate the large signal solution, and then performing a time domain noise simulation to calculate the time varying noise variance of the periodic waveform at the output. We already saw an example for such a noise analysis in Section 5.10.2 for a CMOS inverter. Then, using the time varying noise variance obtained from noise analysis and the slew rate of the large signal waveform at the transitions, one can characterize the timing jitter of the transitions. The timing jitter process J_{fd} contributed by an FD can be quite accurately modeled as a discrete-time, zero-mean, WSS *white* Gaussian process, which means that $J_{fd}[k]$s are *uncorrelated* random variables for different k. For comparison, recall that the timing jitter J_{vco} contributed by the VCO is modeled as a random walk process, i.e. $J_{vco}[k]$s are *correlated* random variables for different k.

For a complete second-order probabilistic characterization of the WSS white discrete time process J_{fd}, all we need is the variance of the random variable $J_{fd}[k]$, which does not depend on k since J_{fd} is WSS. The variance of $J_{fd}[k]$ can be calculated using the data from the time domain noise analysis of the FD. It is calculated by dividing the noise variance at the transitions of the output of the FD by the square of the slew rate (i.e. time derivative) of the large signal periodic waveform at the same transition times. As a matter of fact, we define the timing jitter at the output of the FD just like we did it for an oscillator in Section 6.2 and use the same characterization algorithm that is based on using the time domain noise simulation of Chapter 5. Since the FD is a nonautonomous system, the noise variance waveform at the output is a periodic steady-state one. Recall that the noise variance waveform for free running oscillators has a linear ramp envelope and grows without bound.

7.2.3 Phase-frequency detector

The phase-frequency detector (PFD) is modeled as a digital state machine, as in [61]. Depending on its design, it is sensitive to either the low-to-high or high-to-low transitions of its two inputs. PFD has two digital outputs which control the two switches for the two current sources in the charge pump. Depending on its design, the outputs are active, i.e. the current sources are on, when the outputs are either in the high or low state. A standard PFD can be in only three of the possible four states of the outputs, i.e. the state where both of the outputs are active is not allowed. The state machine model for a PFD, which is sensitive to low-to-high transitions, is shown in Figure 7.3. Every time there is a transition event at one of the inputs of the PFD, the transition type (TRANSITION which is either LOW-TO-HIGH or HIGH-TO-LOW) and the input (INPUT, set to REF for the reference oscillator, and to VCO for the output of the FD inside the feedback loop) where transition has occurred is passed to the code in Figure 7.3 to determine the next state of the PFD. The NEUTRAL state for the PFD is the state where both of its outputs are inactive. The UP and DOWN states refer to the states where one of the outputs is active and the other one is inactive. In the model of Figure 7.3, when there is a LOW-TO-HIGH transition at the REF input of the PFD, the state of the

```
if   TRANSITION == LOW-TO-HIGH      {
       if PFDSTATE == NEUTRAL          {
          if INPUT == REF              PFDSTATE = UP;
          elseif INPUT == VCO          PFDSTATE = DOWN;
       }
       elseif PFDSTATE == UP           {
          if INPUT == REF              PFDSTATE = UP;
          elseif INPUT == VCO          PFDSTATE = NEUTRAL;
       }
       elseif PFDSTATE == DOWN         {
          if INPUT == REF              PFDSTATE = NEUTRAL;
          elseif INPUT == VCO          PFDSTATE = DOWN;
       }
}
```

Figure 7.3. State machine model of a standard PFD

PFD changes from DOWN to NEUTRAL or NEUTRAL to UP. If it is already in the UP state, it stays there. On the other hand, when there is a LOW-TO-HIGH transition at the VCO input of the PFD, the state of the PFD changes from NEUTRAL to DOWN or UP to NEUTRAL. If it is already in the DOWN state, it stays there.

The PFD model we just described assumes that the effect of the transitions at its inputs appears at its outputs instantaneously, without any delay. This is, of course, not possible for a practical PFD. There will be a finite delay before the effect of the changes at its inputs are propagated to its outputs. This finite delay renders the PFD insensitive to the transitions at its inputs that are separated with a time difference that is less than the propagation delay. This results in, what is commonly referred to as, a *dead-zone* for the PFD. For instance, if there are closely spaced transitions (with a separation that is less than the propagation delay of the PFD) at the input of the PFD which require it to change its state from NEUTRAL to UP and back to NEUTRAL again, the PFD will not be able to respond to these transitions, and its state will remain at NEUTRAL.

To model the dead-zone, we form an *event queue* for the PFD during simulation. Every time a transition occurs at one of the inputs of the PFD, it is placed in the event queue. Then, the event queue is processed to check for consecutive transitions which require the PFD to switch its state to create pulses at its outputs which have widths that are smaller than the propagation delay. If such consecutive transitions are detected, they are removed from the event queue. The consecutive transitions that create pulses with widths larger than the propagation delay remain in the queue, and they are passed to the code in Figure 7.3 at their scheduled times to determine the next state of the PFD. When transitions are placed in the event queue, they are scheduled for evaluation after one propagation delay time.

PFD can not create pulses at its outputs with a width that is less than the propagation delay, but it can create pulses at both of the outputs which have widths that are larger than the propagation delay. The difference of the widths of the pulses at the outputs can be *smaller* than the propagation delay. Thus, one can obtain an *effective* pulse at one of the outputs with a width that is less than the propagation delay. This is the basic idea behind the *alive-zone* PFD described in [62]. The alive-zone PFD introduces a reset delay into a standard edge-sensitive PFD and creates minimum duration pulses for each phase comparison at both of its outputs, and hence enables the fourth state where both of the outputs are active. In this state, both of the current sources of the charge pump are on. The positive and negative charge pump currents both deliver a charge greater than zero at each phase comparison. Ideally, with zero phase error, the net charge pumped sums to zero. The alive-zone PFD is also modeled for simulation using a simple event queue (which has only one event at a time) to realize the minimum width pulses at each phase comparison. The width of the pulses the alive-zone PFD creates at its outputs (even when there is zero phase error) is a design parameter, and, obviously, has to be greater than the propagation delay.

PFD also contributes to the timing jitter of the signal at its output. The timing jitter characterization for the PFD can also be done using time domain noise simulation as it was described for the FD.

7.2.4 Charge pump and the loop filter

The charge pump (CP) usually consists of two current sources which are turned on and off with the two switches that are controlled by the outputs of the PFD. The loop filter (LF) is usually a simple passive RLC network, but LFs with active components are also used. We will first present a general model for the CP and the LF, and then describe the model of a very simple CP-LF configuration within the framework of the general model as an example. The CP and the LF are modeled with

$$\begin{aligned} \dot{\mathbf{x}} &= \mathcal{G}(\mathbf{x}, S_{up}, S_{down}) \\ v_c &= \mathbf{c}^T \mathbf{x} \end{aligned} \qquad (7.23)$$

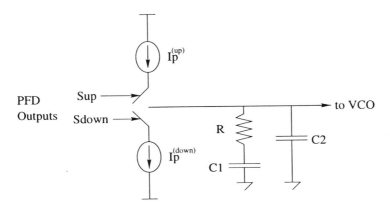

Figure 7.4. Simple charge pump and second-order loop filter

where S_{up} and S_{down} are the two outputs of the PFD, and they are equal to 1 when they are active, and 0 when inactive. Recall that for a standard PFD with a dead-zone, the state

$$S_{up} = 1 \quad S_{down} = 1 \quad (7.24)$$

is not allowed. $\mathbf{x} \in \mathbb{R}^n$ in (7.23) is a vector of state variables which represent the states of the CP and the LF. $\mathcal{G} : \mathbb{R}^n \times \{0,1\} \times \{0,1\} \to \mathbb{R}^N$ is a nonlinear map that describes the dynamics for the states of the CP and the LF. v_c is the output of the LF that controls the frequency of the VCO.

Let us now consider a simple example. A CP with ideal current sources that are controlled with ideal switches, and a simple second-order LF are shown in Figure 7.4. We choose the capacitor voltages v_1 and v_2 as the state variables. Then, the equations that describe the dynamics of the CP and the LF are given by

$$\begin{aligned} \dot{v}_1 &= \tfrac{1}{RC_1} v_2 - \tfrac{1}{RC_1} v_1 \\ \dot{v}_2 &= \tfrac{(I_p^{(up)} S_{up} - I_p^{(down)} S_{down})}{C_2} - \tfrac{1}{RC_2} v_2 + \tfrac{1}{RC_2} v_1 \end{aligned} \quad (7.25)$$

The control voltage for the VCO is the voltage across C_2. Hence,

$$v_c = [0 \; 1][v_1 \; v_2]^T. \quad (7.26)$$

The model represented by (7.23) is general enough, so one can include all kinds of nonidealities in this model. For practical charge pumps, the current sources can be modeled as nonlinear ones with finite output impedances. The loading impedance of the VCO on the CP can also be included in this model.

The output of the LF controls the frequency of the VCO, hence it is an extremely important node in the circuit. Any kind of noise on this node will directly translate into spurious tones and phase noise at the output of the VCO. Ideally, this node should not have a resistive path to ground. For instance, this is the case for the simple CP-LF configuration in Figure 7.4. When the loop is in lock, this node will keep its charge and hold the VCO at a constant frequency. Loading impedance of the VCO on the LF, or the finite output impedances of the current sources will cause charge leakage from this node. Then, the feedback loop will try to compensate for this leakage, and the PFD will turn on the CP current sources so that the leaked charge can be replaced. If the leakage is severe, the CP current sources will turn on and off at every phase comparison at the PFD. Hence, it is extremely important to accurately model the leakage at the input and the output of the LF (e.g. finite output impedances for the CP current sources, loading impedance of the VCO, etc.) so that the spurious tones at the output of the VCO can be estimated accurately by simulating the PLL model.

7.3 Behavioral Simulation Algorithm

We described the models for all of the components of the PLL in Figure 7.1. We will now put these models together to form the simulation model for the whole PLL:

$$\dot{\theta}_{ref}(t) = 2\pi f_{ref}$$
$$\theta_{ref}(t_k^{(ref)}) = k\pi + \phi_{ref}[k] + \phi_{fd}[k] + \phi_{pfd}[k] \tag{7.27}$$

$$\dot{\theta}_{fd}(t) = \frac{2\pi g(v_c)}{N}$$
$$\theta_{fd}(t_k^{(fd)}) = k\pi + \frac{\phi_{vco}[k]}{N} \tag{7.28}$$

$$\dot{\mathbf{x}} = \mathcal{G}(\mathbf{x}, S_{up}, S_{down})$$
$$v_c = \mathbf{c}^T \mathbf{x} \tag{7.29}$$

where $k = \{0, 1, 2, 3, \ldots\}$.

(7.27) is the model of the reference oscillator, but we included the phase noise contributions of the FD and the PFD, i.e. ϕ_{fd} and ϕ_{pfd}, in the model of the reference oscillator. This is based on the results we obtained from the simple LTI analysis in Section 7.1. The phase noise contributed by the FD and the PFD is *indistinguishable* from the phase noise of the reference oscillator when we evaluate their effects on the phase noise of the oscillation signal produced by the PLL at the output of the VCO. ϕ_{ref} for the reference oscillator is modeled as a random walk process, on the other hand, ϕ_{fd} and ϕ_{pfd} for the FD and the PFD are modeled as WSS white discrete-time processes.

(7.28) models the VCO signal and the phase noise contribution of the VCO at the output of the FD. Alternatively, we could have modeled the VCO signal at the output of the VCO itself. The node between the VCO and the FD is a high frequency node. By modeling the VCO signal at the output of the FD, we effectively eliminate this high frequency node which makes N transitions for every one transition of the FD output. With almost no loss of accuracy, eliminating this high frequency node increases the efficiency of the simulation considerably, especially when N is a very large integer.

(7.29) models the CP and the LF. The state machine model for the PFD was described in Section 7.2. When the output of the reference oscillator and the output of the FD make transitions at times $t_k^{(ref)}$ and $t_k^{(fd)}$ determined by (7.27) and (7.28), these transition events are scheduled in the event queue for the PFD.

7.3.1 Numerical integration with threshold crossing detection

To simulate the PLL model, we solve the following system of differential equations in time domain using numerical integration

$$\begin{aligned} \dot{\theta}_{ref}(t) &= 2\pi f_{ref} \\ \dot{\theta}_{fd}(t) &= \frac{2\pi g(\mathbf{c}^T \mathbf{x})}{N} \\ \dot{\mathbf{x}} &= \mathcal{G}(\mathbf{x}, S_{up}, S_{down}) \end{aligned} \quad (7.30)$$

where we replaced v_c with $\mathbf{c}^T \mathbf{x}$. At the time instants $t_k^{(ref)}$ and $t_k^{(fd)}$, when θ_{ref} and θ_{fd} cross the threshold values

$$\theta_{threshold}^{(ref)}[k] = k\pi + \phi_{ref}[k] + \phi_{fd}[k] + \phi_{pfd}[k] \quad (7.31)$$

and

$$\theta_{threshold}^{(fd)}[k] = k\pi + \frac{\phi_{vco}[k]}{N} \quad (7.32)$$

respectively, transition events need to be scheduled in the PFD event queue. Hence, the numerical integration algorithm has to detect these threshold crossings. Accurate estimation of the timing of these threshold crossings is crucial, because we will characterize the spurious tones and the timing jitter for the PLL output based on the timings of the transitions calculated during the simulation. At a time point during the numerical integration of (7.30), if a transition is detected in between the previous time point and current one, the integration algorithm should track back and estimate the exact time point where the transition occurred more accurately, as illustrated in Figure 7.5.

The first thing that comes to mind to estimate the timings of the transitions is to use *interpolation*. With this scheme, transition times are estimated

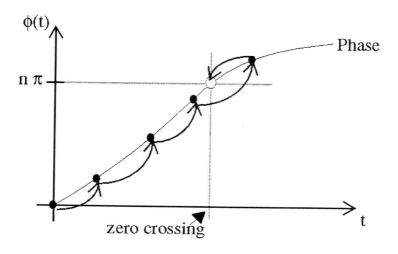

Figure 7.5. Integration algorithm with threshold crossing detection

based on a polynomial interpolation using the information from the current and previous time points. Linear multi-step methods for numerical integration actually calculate the solution of the differential equation at discrete time points by fitting polynomials. Hence, to estimate the timing of the transition times, one can and should use the same polynomial that was used by the numerical integration scheme to calculate the solution at the current time point. For instance, if backward Euler was the scheme in calculating the solution at the current time point, one uses a first-order polynomial to interpolate the timing of the detected threshold crossing. Let a transition event be detected between the time points t and $t + h$, where h is the current time step size. Then, the timing of the transition event is estimated by

$$t_{tr} = t + \left[\frac{\theta_{threshold} - \theta(t)}{\theta(t+h) - \theta(t)}\right] h \qquad (7.33)$$

where $\theta_{threshold}$ is the threshold value that was crossed.

Alternatively, one can estimate the timing of the transitions using an *iterative method*. Actually, calculating the transition time can be interpreted as the calculation of the solution of a nonlinear algebraic equation with one variable, i.e. we would like to calculate the time t that satisfies

$$\theta(t) - \theta_{threshold} = 0. \qquad (7.34)$$

Hence, to estimate the transition time, one can use any algorithm for finding the roots of a nonlinear algebraic equation. Obviously, $\theta(t)$ is not available

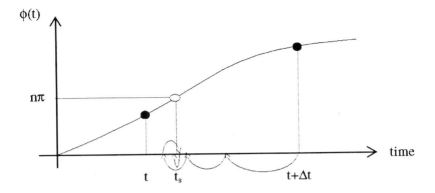

Figure 7.6. Iterative solution for transition times

analytically as a function of t, but it can be evaluated numerically using the numerical integration scheme. A brute-force method, which works quite well, to calculate the root of (7.34) is to use a bisection search, which is illustrated in Figure 7.6. Recall that the phase $\theta(t)$ is a monotonically increasing function of t. If at time $t + h$, the phase $\theta(t)$ is larger than the threshold $\theta_{threshold}$ while at t it was smaller, then a transition event has occurred in between. To calculate the transition time, we divide the present time step by 2. Then, we evaluate $\theta(t + h/2)$. If this is still to the right of the transition time, we repeat the bisection procedure. If it is to the left, then we return the control to the integration algorithm. To prevent the algorithm from taking too much of a time step, we limit the maximum step size to a quarter of the last time step. Other iterative algorithms can also be used to calculate the root of (7.34). One can first use the interpolation scheme to calculate an initial starting point for the iterative algorithm that is used to calculate the root of (7.34). The iterative algorithm, e.g. bisection search, is terminated when the *tolerance* specified for calculating the timing of the transition times is reached. Tolerance for the calculating the transitions is set to a small enough value, so that the phase noise/timing jitter that will be estimated for the PLL output will not be corrupted by the numerical noise due to the errors in calculating the transition times.

Threshold crossings of θ_{ref} and θ_{fd} cause events at the inputs of the PFD. The events at the output of the PFD (coming out of its event queue) change the state of the PFD, i.e. they set or reset the two outputs S_{up} and S_{down}. Every time an event is placed in the event queue for the PFD, a *break point* in time is scheduled, so that the numerical integration algorithm will place a time point at the time instants where PFD outputs may switch their values. After each break point, a "small" time step is taken and the numerical integration scheme is reset to backward Euler (i.e to order 1) if a variable order numerical scheme is being used.

Some commercial circuit simulators, which support behavioral modeling for analog and mixed-signal circuits with a behavioral analog hardware description language, provide constructs for the detection of thresholds crossings and to estimate the timings for them. They also provide constructs for scheduling break points in time.

7.3.2 Simulating the timing jitter in numerical integration

Timing jitter of the reference oscillator, VCO, FD and the PFD is introduced into the time domain noise simulation as in (7.27) and (7.28) through the threshold values which indirectly determine the transition times. The threshold value for the reference oscillator and the output of the FD are discrete-time stochastic processes given by (7.31) and (7.32). The phase noise processes for the reference oscillator and the VCO are modeled as random walk processes,[7] and the phase noise processes for the FD and the PFD are modeled as white WSS discrete time processes. In time domain noise simulation, they are realized by using a Gaussian random number generator. To realize a random walk phase noise process with a Gaussian random number generator, we use the construction of a random walk process given by (6.32) and (6.38) that was discussed in Section 6.2.2. With this construction, we can generate a random walk process by using uncorrelated identically distributed Gaussian random variables. Hence, all we need is a standard Gaussian random number generator. The realization of a white WSS Gaussian phase noise process is straightforward: a sequence of uncorrelated Gaussian random variables generated from the same distribution.

7.3.3 Acquisition detection and the simulation output

The simulation algorithm has the following three components, which were all described above:

- Numerical integration algorithm with threshold crossing detection and break points.

- PFD event queue scheduling and evaluation.

- Realization of the phase noise/timing jitter processes using a Gaussian random number generator.

During the simulation, events are placed into the PFD event queue at the time instants of the threshold crossings detected by the integration algorithm. At the threshold crossing instants, break points are scheduled for the time points where the events coming out of the PFD queue need to be evaluated.

When the simulation is started at $t = 0$, the phase noise sources are turned off. The PLL is simulated without the phase noise sources, i.e. with deterministic threshold values set to $k\pi$ for the kth transition, till the VCO locks onto the reference signal. Acquisition of the reference signal is detected automatically by observing $v_c = \mathbf{c}^T\mathbf{x}$ which controls the frequency of the VCO. Even though the phase noise sources are turned off during acquisition, there may be systematic error sources in the PLL (e.g. leakage from the node at the output of the LF, dead-zone of the PFD, etc.) which will cause steady-state fluctuations in $v_c = \mathbf{c}^T\mathbf{x}$ even after the PLL has locked onto the reference signal. Thus, the algorithm that automatically detects acquisition uses a tolerance parameter so that it can declare the PLL as locked even when there are steady-state fluctuations in $v_c = \mathbf{c}^T\mathbf{x}$. On the other hand, the PLL may never lock onto the reference signal, for instance if the feedback loop is not stable. Acquisition detection algorithm aborts the simulation if the PLL can not lock onto the reference signal in a specified amount of time. $v_c = \mathbf{c}^T\mathbf{x}$ and any other state variable in $\mathbf{x}$ are dumped at every time point during the acquisition phase, so that they can be plotted to observe the acquisition behavior of the PLL.

If the algorithm that automatically detects acquisition declares the PLL as locked, then the phase noise sources are turned on. The PLL, which is in lock, is simulated for a period of time[8] in the presence of the phase noise sources of the reference oscillator, VCO, FD and the PFD, and systematic error sources. During this simulation, the control voltage $v_c = \mathbf{c}^T\mathbf{x}$ as a function of time is dumped for postprocessing. The timings of the threshold crossings detected at the output of the FD by the numerical integration algorithm are also dumped for postprocessing. The postprocessing of the simulation output to evaluate several performance measures for the PLL will be described in the next section.

7.4 Post Processing for Spurious Tones and Timing Jitter/Phase Noise

The acquisition behavior of the PLL is observed by plotting the dumped control voltage $v_c = \mathbf{c}^T\mathbf{x}$ during the time period when the PLL is acquiring the reference signal. Observing the steady-state fluctuations in the waveforms of $v_c = \mathbf{c}^T\mathbf{x}$ and other state variables in $\mathbf{x}$ after the PLL has locked can provide crucial information about the systematic error sources in the PLL that cause spurious tones at the output. For instance, by observing these waveforms, one can observe the severity of the leakage from the nodes at the input and the output of the LF. In the next two subsections, we will describe the post processing techniques used to estimate the spurious tones and the timing jitter/phase noise of the oscillation signal generated by the PLL at the output of the VCO.

7.4.1 Spurious tones

The frequency of the signal at the output of the VCO is controlled by the control voltage $v_c = \mathbf{c}^T \mathbf{x}$, i.e.

$$f_{out}(t) = g(v_c(t)). \tag{7.35}$$

When the PLL is locked onto a reference signal with frequency f_{ref}, at steady-state, the VCO frequency $f_{out}(t) = g(v_c(t))$ will have a mean value that is equal to Nf_{ref}. However $f_{out}(t)$ will exhibit fluctuations around this mean value due to the phase noise contributed by the loop components and other systematic error sources. During the time domain noise simulation of the PLL, we dumped the steady-state waveform for $v_c(t)$. In general, the steady-state waveform for $v_c(t)$ will be neither periodic nor quasi-periodic. It will be a *chaotic* steady-state waveform. However the power in the fluctuations of $v_c(t)$ are usually concentrated at a set of frequencies. For instance, $v_c(t)$ may contain distinguishable spectral components at the reference oscillator frequency and its several harmonics. One can usually identify these strong spectral components by just looking at the waveform of $v_c(t)$. By performing Fourier analysis (using an appropriate numerical technique such as FFT, almost periodic transform, etc. [29, 63]) on the steady-state waveform of $v_c(t)$ obtained from time domain noise simulation, one can identify and quantify these spectral components. At this point, we assume that the steady-state waveform of $v_c(t)$ can be approximated as a quasi-periodic waveform by identifying the strong spectral components, i.e.

$$v_c(t) \approx \sum_{m=-M}^{M} a_m \exp\left(j2\pi f_m t\right) \tag{7.36}$$

where $f_{-m} = -f_m$, $m = \{0, 1, 2, \ldots\}$ is a set of frequencies at which the steady-state waveform $v_c(t)$ has significant power. The frequency of the VCO, f_{out}, is related to the control voltage v_c through a nonlinear map $g(.)$. Hence, in general, f_{out} will have spectral components at the frequencies f_m as well as their harmonics. However the fluctuations in v_c are usually small around the mean value $g^{-1}(Nf_{ref})$. Thus, the spectral components of f_{out} at the harmonics of the frequencies f_m will be negligible. Hence, one can linearize the nonlinear map $g(.)$ around the mean value $g^{-1}(Nf_{ref})$ of v_c to calculate the spectral components of f_{out}. At this point, we assume that f_{out} has the following approximate quasi-periodic representation

$$f_{out}(t) \approx \sum_{m=-M}^{M} b_m \exp\left(j2\pi f_m t\right) \tag{7.37}$$

where $f_0 = 0$, and hence the zeroth order term is given by

$$b_0 = N f_{ref}. \tag{7.38}$$

Let us also assume that the free running VCO generates the following periodic waveform[9] with fundamental frequency $N f_{ref}$ at its output

$$V_{osc}(t) = \sum_{l=-L}^{L} c_l \exp\left(j2\pi l N f_{ref} t\right) \tag{7.39}$$

which is expressed as a Fourier series with coefficients c_l. (7.39) is the waveform that is generated by a free running VCO with its control voltage set to the constant $g^{-1}(N f_{ref})$. To approximate the waveform at the output of the VCO in the PLL with its control voltage set to (7.36), we replace[10] $N f_{ref} t$ in (7.39) with

$$\int^t f_{out}(\tau)d\tau \approx N f_{ref} t + \sum_{\substack{m=-M \\ m \neq 0}}^{M} \frac{b_m}{j2\pi f_m} \exp\left(j2\pi f_m t\right) \tag{7.40}$$

to obtain

$$V_{osc}(t) \approx \sum_{l=-L}^{L} c_l \exp\left[j2\pi l \left(N f_{ref} t + \sum_{\substack{m=-M \\ m \neq 0}}^{M} \frac{b_m}{j2\pi f_m} \exp\left(j2\pi f_m t\right)\right)\right]. \tag{7.41}$$

The waveform in (7.41) can be interpreted as a *frequency modulated* periodic carrier. The spectrum of the waveform in (7.41) has nonzero components at frequencies $lN f_{ref} \pm k f_m$ for $l = 0, 1, 2, \ldots, L$ and $k = 0, 1, 2, \ldots$, for each frequency modulation component at frequency f_m, $m \in \{1, 2, \ldots, M\}$. One is usually most concerned with the components of (7.41) at frequencies $N f_{ref} \pm k f_m$ for $k = 0, 1, 2, \ldots$, for each f_m. These frequency components are usually referred to as the *spurious tones* in the output of the PLL. One can estimate the magnitudes of the spurious tones in (7.41) by using techniques from frequency modulation theory, i.e. Bessel functions. We will not discuss the details of how one can do this, since it is covered in any textbook that discusses frequency modulation in communications.

7.4.2 Timing jitter/phase noise

The spurious tones in the VCO output are caused by the systematic error sources in the PLL, which show up as periodic or quasi-periodic fluctuations at a number of frequencies in the control voltage v_c dumped from the simulation. With only systematic error sources, the spectrum of the VCO output would contain components at discrete frequencies. These are the spurious tones we calculated in the previous section.

The phase noise sources of the VCO, FD, reference oscillator, and the PFD will also effect the output of the VCO. To characterize the phase noise of the VCO output, we use the sequence of transition times, i.e. threshold crossing times, at the output of the FD, which were saved during the simulation of the PLL after it acquired the reference oscillaton signal. If there were no error sources in the PLL, i.e. no phase noise contribution from any of the components, and also no systematic error sources, then the transition times dumped from the simulator would be *evenly* spaced in time with a separation that is equal to the half of the period of oscillation, i.e. with a separation that is equal to

$$\frac{1}{2f_{ref}}. \quad (7.42)$$

Because of the noise sources in the PLL, these transition times will not be evenly spaced in time, and they will, in general, be random variables. Let γ be the discrete-time stochastic process that represents these random transition times, and let ζ be the discrete-time stochastic process that represents the deviation of these random transition times from the ideal ones that are evenly spaced with separation as in (7.42). ζ can be obtained from γ by

$$\zeta[k] = \gamma[k] - \left(k\frac{1}{2f_{ref}} + t_{synch}\right) \qquad k = 0, 1, 2, \ldots \quad (7.43)$$

t_{synch} in (7.43) is the time instant for the ideal transition that corresponds to $k = 0$, and it is unknown. Even though we have a sample path of γ from the simulation, we can not calculate the sample path for ζ, since t_{synch} is unknown [38]. γ satisfies

$$\mathrm{E}\left[\gamma[k]\right] = k\frac{1}{2f_{ref}} + t_{synch} \qquad k = 0, 1, 2, \ldots \quad (7.44)$$

Hence, ζ is a zero mean process.

We would like to calculate a second-order probabilistic characterization of the timing jitter/phase noise of the oscillation waveform that is represented by the transition time deviations ζ. We obtain a stochastic process J from ζ using *self-referencing* [38] as follows:

$$J[k] = \zeta[k] - \zeta[0] \qquad k = 0, 1, 2, \ldots \tag{7.45}$$

for $k = 0, 1, 2, \ldots$. Note that

$$J[0] = 0. \tag{7.46}$$

Even though t_{synch} is unknown, we can calculate the sample path for J from the sample path of γ we obtained from simulation. From (7.43) and (7.45), J in terms of γ is given by

$$J[k] = \gamma[k] - \left(k \frac{1}{2f_{ref}} + \gamma[0]\right) \qquad k = 0, 1, 2, \ldots \tag{7.47}$$

For the PLL, the discrete-time process ζ that represents the transition time deviations is a WSS process in steady-state (assuming that the reference signal is noiseless).[11] The way we obtained the timing jitter process J from ζ is called *self-referencing*, because we calculate the jitter $J[k]$ in (7.45) at the transitions for $k = 1, 2, \ldots$ with respect to the *reference* transition at $k = 0$.[12] Even though ζ is a WSS process, the self-referenced jitter process J defined by (7.45) is not WSS, but it is *asymptotically* WSS.

7.4.2.1 Variance of the timing jitter of transitions.

Using the dumped transition times (i.e. a sample path for γ) from the simulation of the PLL, we can generate a *sample path* for the self-referenced timing jitter process J using (7.47). In general, we need to compile an *ensemble* of sample paths for J so that we can calculate ensemble averages, i.e. expectations, to estimate its second-order probabilistic characteristics. However, if J satisfies some *ergodicity* properties, then, we need only a *single* sample path to characterize its probabilistic characteristics by calculating expectations using *time averages* instead of ensemble averages. Next, we explore the ergodicity properties of the self-referenced timing jitter process J.

Let R_ζ be the autocorrelation function of the WSS process ζ. It can be shown that R_ζ is in the form

$$R_\zeta[m] = \rho z[m] \tag{7.48}$$

where ρ is the variance of the zero mean, WSS process ζ, i.e.

$$\rho = \mathrm{E}\left[\zeta[k]^2\right] \tag{7.49}$$

and $z[m]$ satisfies

$$z[m] \to 0 \quad \text{as} \quad |m| \to \infty \tag{7.50}$$

and
$$z[0] = 1. \tag{7.51}$$

Thus, $z[m]$ is the normalized autocorrelation function: It is equal to 1 at $m = 0$ and decays to 0 as $|m|$ is increased. Actually, $z[m]$ will be approximately 0 for $|m| > M$ for some positive integer M. M is roughly determined by the PLL feedback loop time constant, i.e. $M \frac{1}{2f_{ref}}$ is roughly equal to the loop time constant [38]. The particular shape of $z[m]$ is not important for our discussion here. We will only use the fact that it becomes negligible for $|m| > M$.

Given the autocorrelation function of ζ, we can calculate the autocorrelation function R_J of the self-referenced jitter process J defined by (7.45) as follows:

$$\begin{align}
R_J[k,m] &= \mathsf{E}\left[J[k]J[m]\right] \tag{7.52} \\
&= \mathsf{E}\left[(\zeta[k] - \zeta[0])(\zeta[m] - \zeta[0])\right] \tag{7.53} \\
&= R_\zeta[k-m] - R_\zeta[k] - R_\zeta[m] + R_\zeta[0] \tag{7.54} \\
&= \rho(z[k-m] - z[k] - z[m] + 1). \tag{7.55}
\end{align}$$

In particular, the variance of J is given by

$$\mathsf{E}\left[J[k]^2\right] = R_J[k,k] = 2\rho(1 - z[k]). \tag{7.56}$$

Note that $\mathsf{E}\left[J[0]^2\right] = 0$, i.e. the variance of the self-referenced timing jitter process in (7.56) is zero at the reference transition represented by $k = 0$, which also follows from (7.46). It reaches the steady-state value of 2ρ for large k. Recall from Chapter 6 that the variance of the timing jitter process for a free running oscillator increases linearly without bound, and does not reach a steady-state value.

Now, let S be the length of the sample path of the transition times γ we obtained from simulation. Let $\gamma_{sp}[k]$, $k = 0, 1, \ldots, S$ denote this sample path. Then, we can calculate the sample path $J_{sp}[k]$, $k = 0, 1, \ldots, S$ for the self-referenced jitter process J by using (7.47). We assume that $M \ll S$. Recall that

$$z[k] \approx 0 \quad \text{for} \quad |k| > M. \tag{7.57}$$

Hence,

$$\mathsf{E}\left[J[k]^2\right] \approx 2\rho \quad \text{for} \quad |k| > M. \tag{7.58}$$

Thus, J is an asymptotically WSS process. We ensure that $M \ll S$ is satisfied so that the sample path of J we generated represents its *steady-state* characteristics.

We would like to estimate the probabilistic characteristics of J using the sample path $J_{sp}[k]$, $k = 0, 1, \ldots, S$. First, we calculate the *sample mean* that is given by

$$m_{sp} = \frac{1}{S} \sum_{i=0}^{S} J_{sp}[i]. \qquad (7.59)$$

Obviously, the sample mean in (7.59) is an *unbiased* estimator for the mean of J, i.e.

$$\mathrm{E}\left[m_{sp}\right] = 0 = \mathrm{E}\left[J[k]\right]. \qquad (7.60)$$

Next, we calculate the *mean-square* error of the sample mean as follows

$$\mathrm{E}\left[(m_{sp} - \mathrm{E}\left[m_{sp}\right])^2\right] = \mathrm{E}\left[m_{sp}^2\right] \qquad (7.61)$$

$$= \frac{1}{S^2} \sum_{i=0}^{S} \sum_{j=0}^{S} \mathrm{E}\left[J_{sp}[i] J_{sp}[j]\right] \qquad (7.62)$$

$$\approx \rho \qquad (7.63)$$

where we used (7.55), (7.57) and $M \ll S$. Thus, (7.63) tells us that the mean-square error in estimating the mean of J using a sample path of length S *does not* decay to 0 as the length S of the sample path is increased. The sample mean m_{sp} can be interpreted as a random variable with mean 0 and variance ρ. Thus,

Remark 7.1 *The self-referenced jitter process J does not possess mean-square ergodicity of the mean.*

We observed this in practice by calculating the sample means of an *ensemble* of sample paths of the jitter process that was generated from the simulation of a PLL.

Next, we calculate the *sample variance* V_{sp} for J using the sample path $J_{sp}[k]$, $k = 0, 1, \ldots, S$ as follows

$$V_{sp} = \frac{1}{S} \sum_{i=0}^{S} (J_{sp}[i] - m_{sp})^2 \qquad (7.64)$$

where the sample mean m_{sp} is given by (7.59). By taking the expectation of both sides of (7.64), we obtain

$$\mathrm{E}\left[V_{sp}\right] = \frac{1}{S} \sum_{i=0}^{S} \left(\mathrm{E}\left[J_{sp}[i]^2\right] + \mathrm{E}\left[m_{sp}^2\right] - 2\mathrm{E}\left[J_{sp}[i] m_{sp}\right]\right). \qquad (7.65)$$

It can be shown that

$$\mathrm{E}\left[J_{sp}[i] m_{sp}\right] \approx \rho \qquad (7.66)$$

by using (7.55), (7.57) and $M \ll S$. Then, we use $M \ll S$, (7.58), (7.63) and (7.66) in (7.65) to obtain

$$E[V_{sp}] \approx \frac{1}{S} \sum_{i=0}^{S} (2\rho + \rho - 2\rho) \qquad (7.67)$$

$$\approx \rho. \qquad (7.68)$$

Hence,

Remark 7.2 *The sample variance V_{sp} is an* unbiased *estimator for*

$$\rho = E\left[\zeta[k]^2\right]. \qquad (7.69)$$

Next, we calculate the *mean-square* error of the sample variance:

$$E\left[(V_{sp} - E[V_{sp}])^2\right] = E\left[V_{sp}^2\right] - E[V_{sp}]^2 \qquad (7.70)$$

$$= E\left[V_{sp}^2\right] - \rho^2. \qquad (7.71)$$

Calculation of $E\left[V_{sp}^2\right]$ requires the calculation of the fourth-order moments of J, but we know only the second-order moments of J, i.e. its autocorrelation function. To calculate the fourth-order moments for J, we assume that it is a Gaussian process. Then, the fourth-order moments can be expressed in terms of the second-order moments using Isserlis's formula [5]. Using (7.55), (7.57), $M \ll S$, (7.58), (7.63), (7.66) and Isserlis's formula for calculating the fourth-order moments of J one can show that

$$E\left[V_{sp}^2\right] \approx \rho^2. \qquad (7.72)$$

Then, from (7.71) and (7.72) if follows that

$$E\left[(V_{sp} - E[V - sp])^2\right] \approx 0. \qquad (7.73)$$

Thus,

Remark 7.3 *The mean-square error in estimating ρ goes to zero as we increase the length S of the sample path, when the sample variance V_{sp} defined by (7.64) is used as the estimator.*

We observed this in practice by calculating the sample variances of the sample paths of the jitter process that was generated from the simulation of a PLL.

$$\rho = E\left[\zeta[k]^2\right] \qquad (7.74)$$

is the variance of the timing jitter of the transitions for the oscillation waveform at the output of the FD, and it can be estimated by calculating the sample variance of a *single* sample path for the self-referenced timing jitter process J produced during simulation.

7.4.2.2 Spectral density of phase noise/timing jitter.

The variance ρ of the timing jitter of the transitions is only a partial second-order probabilistic characterization of the self-referenced timing jitter process J. We would also like to calculate an estimate of the autocorrelation function or the *spectral density* of the asymptotically WSS process J.

We calculate the spectral density of the self-referenced timing jitter process J using the time-averaged periodogram method discussed at the end of Section 2.2.11. We apply the time-averaged periodogram method to the sample path $J_{sp}[k]$, $k = 0, 1, \ldots, S$, after we subtract the sample mean from every sample. We already used the same sample path to estimate the variance ρ for the timing jitter. Recall that the sample mean, i.e. the average of the samples in $J_{sp}[k]$, $k = 0, 1, \ldots, S$, can be nonzero. In fact, as derived previously, the sample mean m_{sp} is a random variable with mean 0 and variance ρ. We subtract the sample mean from the sample path to obtain

$$\tilde{J}_{sp}[k] = J_{sp}[k] - m_{sp}, \quad k = 0, 1, \ldots, S. \quad (7.75)$$

We then use the time-averaged periodogram method on the sample path $\tilde{J}_{sp}[k]$, $k = 0, 1, \ldots, S$, to calculate the spectral density of the timing jitter process at the output of the FD. Recall that the timing jitter of the transitions for the oscillation waveform at the input of the FD (i.e. at the output of the VCO) is the same as the timing jitter of the transitions at its output. Hence, J also represents the timing jitter process for the oscillation waveform at the output of the VCO. The phase noise process at the output of the FD is given by

$$2\pi f_{ref} J \quad (7.76)$$

and the phase noise process at the output of the VCO is given by

$$2\pi N f_{ref} J. \quad (7.77)$$

Hence, we can easily calculate the spectral density of the phase noise processes from the spectral density of the timing jitter process J by simply scaling it with $(2\pi f_{ref})^2$ or $(2\pi N f_{ref})^2$.

The timing jitter process J at the output of the FD represents the jitter of the transitions separated by $\frac{1}{2f_{ref}}$. We can interpret J as the output of a sampler with a continuous-time process as its input with a sampling rate that is equal to $2f_{ref}$. So, the spectral densities can be calculated for the frequency range $[-f_{ref}, f_{ref}]$. The phase noise spectral density calculated as such corresponds to the phase noise spectrum that can be measured on a spectrum analyzer as a function of the offset frequency from the carrier, i.e. the fundamental oscillation frequency. The units for the phase noise spectral density is rad^2/Hz. It is usually plotted in decibels, and then the unit is *dBc* which stands for *decibels below carrier*.

If there are systematic error sources in the PLL, the spectral densities calculated for timing jitter/phase noise may exhibit peaks et certain frequencies, i.e. spurious tones. In Section 7.4.1, we described another numerical method to estimate these spurious tones, which was based on performing Fourier analysis on the steady-state VCO control voltage calculated by the simulation algorithm.

7.5 Examples

We will now present examples of PLL behavioral simulation using the techniques presented in this chapter. In particular, we will present examples for the analysis of the acquisition characteristics of PLLs, and timing jitter/phase noise characterization.

The PLL behavioral modeling and simulation techniques presented in this chapter were used in the top-down constraint-driven design of an on-chip clock generator for a video driver system (which was fabricated) [64, 4]. The architecture of this clock generator PLL circuit is the one shown in Figure 7.1, and the the loop filter is as the one shown in Figure 7.4. Some behavioral simulation results will be presented for this PLL, but the reader is referred to [64, 4] for a detailed description of how the behavioral modeling and simulation techniques of this chapter were used in its top-down design.

We used the techniques of this chapter in the bottom-up verification of a frequency synthesizer PLL that was designed to generate the LO signal in a cellular telephony application. We were able to identify the systematic error sources and make predictions for the spurious tone and the phase noise performance of this PLL using the behavioral modeling and simulation methodology. The results we obtained were quite close to the measurements on the fabricated PLL circuit. Thus, the usefulness of the techniques we described in this chapter for the design of PLLs for clock generation and frequency synthesis applications was proven for an industrial design example. Unfortunately, due to the proprietary nature of this project, we will not be able to present the behavioral modeling and simulation results we obtained for this circuit.

7.5.1 Acquisition behavior

Bottom-up verification of a phase-locked system using behavioral simulation is done in two steps:

- Set up the behavioral models for the components. The model parameters are extracted using SPICE and other transistor-level simulation techniques from the transistor-level description of the components.

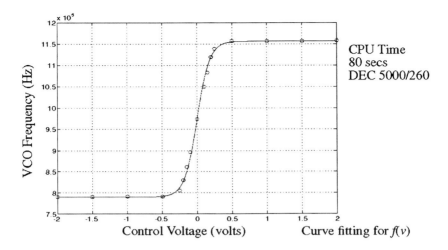

Figure 7.7. Extraction of frequency versus control voltage relationship for the VCO

- Simulate the system in time-domain using behavioral simulation and post-process the simulation output to characterize various performance measures.

To illustrate this procedure for the analysis of the acquisition characteristics, component behavioral models for a bipolar PLL [1] were set up, and behavioral simulation was used to analyze its acquisition characteristics. The VCO of the PLL circuit is the relaxation oscillator we analyzed in Chapter 6. The phase detector is a Gilbert multiplier, and the loop filter is a one pole, passive low pass filter. There is no frequency divider in the feedback loop. Figure 7.7 shows the SPICE domain extraction of the relation between the effective frequency of the VCO and the control voltage. Other model parameters are extracted in a similar way from the transistor level description. Model extraction is done only once for a circuit. Then, created models are used in many behavioral simulations of the PLL. Figure 7.8 shows the response of the modeled PLL to a frequency step at the reference oscillator signal, both SPICE and behavioral simulation. Behavioral simulation is *two* orders of magnitude faster than SPICE.

The acquisition characteristics of the clock generator PLL of [64, 4] was analyzed using behavioral simulation. The waveform shown in Figure 7.9 is the control voltage of the VCO when the PLL is acquiring the reference signal. This simulation was performed to analyze the stability of the feedback loop for the worst case frequency divider ratio. Figure 7.9 shows the waveforms obtained with both behavioral and SPICE simulation, which are almost identical. The

248 NOISE IN NONLINEAR ELECTRONIC CIRCUITS

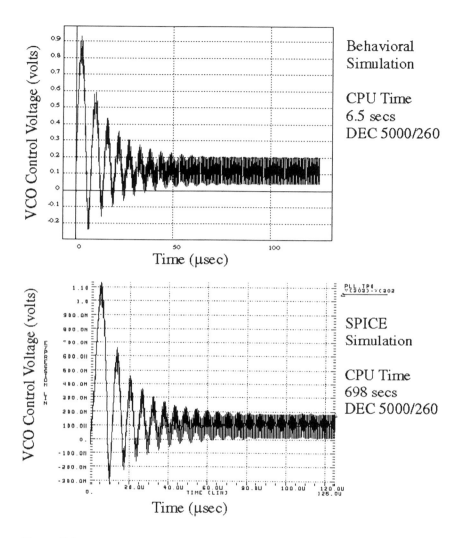

Figure 7.8. Acquisition characteristics using behavioral simulation and SPICE

behavioral simulation took 560 CPU seconds, whereas the full circuit simulation took 20 CPU hours.[13]

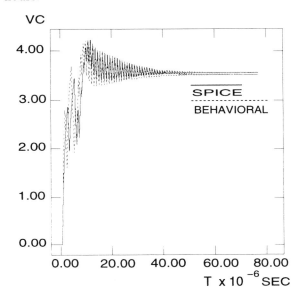

Figure 7.9. Acquisition characteristics of the clock generator PLL

7.5.2 Timing jitter characterization

The role of behavioral simulation in the top-down design of a phase/delay-locked system will be illustrated for the design of a multi-phase clock generator. A PLL with a ring-oscillator VCO or a delay-locked loop (DLL) can be used to generate multi-phase clocks, as in Figure 7.10. These two architectures were compared for their timing jitter performance using behavioral simulation. Figure 7.11 shows the relationship between the rms clock jitter (the square root of ρ that was defined in Section 7.4.2.1) and the percentage delay cell jitter contribution for a given design of both architectures. Ring-oscillator VCO for PLL, as well as the delay line of the DLL, has 5 delay cells. Reference clock frequency is 50 MHz. From Figure 7.11, we conclude that DLL has better timing jitter performance, when compared with a PLL, for fixed percentage delay cell jitter contribution. Then, the relationship between clock jitter and percentage delay cell jitter is used to predict the amount of jitter allowable an a delay cell, given a clock jitter allowance. In this way, the clock jitter constraint is mapped onto the delay cell jitter constraint.

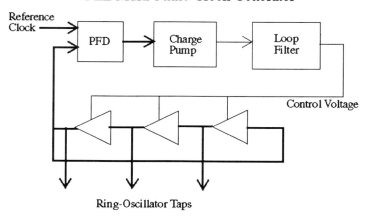

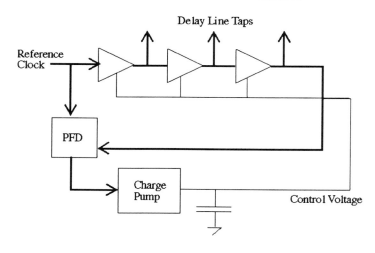

Figure 7.10. PLL and DLL multi-phase clock generators

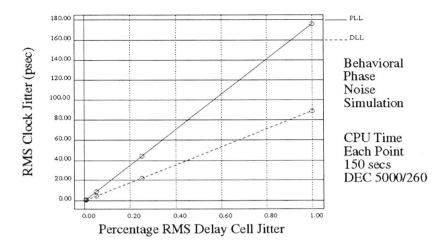

Figure 7.11. Clock jitter for PLL and DLL clock generators

We will now present the behavioral simulation results obtained for the timing jitter characterization for the PLL of [64, 4]. The left plot in Figure 7.12 shows the square root of the variance for the self-referenced timing jitter process for the PLL output. (Please see the discussion in Section 7.4.2.1). The right plot in Figure 7.12 shows the variance of the self-referenced timing jitter process both for the closed-loop PLL output and the free running VCO. The variance for the free running VCO increases linearly without bound. On the other hand, for the closed-loop case, the timing jitter process reaches a steady-state value that is equal to 2ρ. The plots in Figure 7.12 were obtained by generating an *ensemble* of sample paths for the self-referenced timing jitter process from the behavioral simulation of the PLL as described in Section 7.4.2.1. From the simulation results in Figure 7.12, we obtain

$$\sqrt{2\rho} \approx 50 \text{ psecs} \qquad (7.78)$$

for the standard deviation of the timing jitter of the transitions at the output of the VCO. This quantity was also measured for the fabricated PLL circuit[14] to be 65 psecs. For behavioral simulation, only the effect of thermal noise in the ring-oscillator VCO was modeled. Process variations affecting the PLL components, power supply and substrate coupling can cause the measured value deviate from the predicted one. Reflections and coupling from the test board could have also affected the measurements. Considering these factors, the agreement between

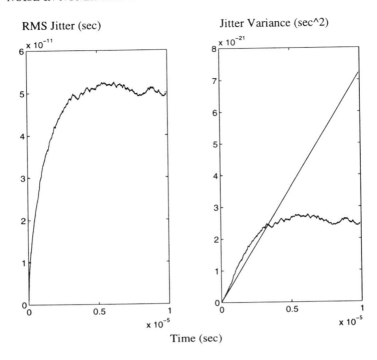

Figure 7.12. Timing jitter characterization of the clock generator PLL

the measured result and the one predicted by behavioral simulation is quite satisfactory [64, 4].

7.5.3 Phase noise spectrum

Now, we illustrate the phase noise spectrum characterization of PLLs with behavioral simulations of a simple PLL circuit shown in Figure 7.13. For the first example, we used an ideal PFD model, a white WSS phase noise source to model the timing jitter contribution at the reference signal, and a random walk phase noise (uncorrelated cycle-to-cycle jitter) model for the open-loop VCO. Figure 7.14 shows the spectrums for the open-loop VCO phase noise, the phase noise source at the reference, and the phase noise of the closed-loop PLL output. Figure 7.15 is an histogram of the sampled phase noise for the closed-loop PLL output. As expected, at low frequencies (below the loop bandwidth) closed-loop phase noise is dominated by the phase noise source at the reference, and at high frequencies by the open-loop VCO phase noise. Practical PFDs with tri-state outputs often have a dead zone region where the output is insensitive to small

BEHAVIORAL MODELING AND SIMULATION OF PHASE-LOCKED LOOPS 253

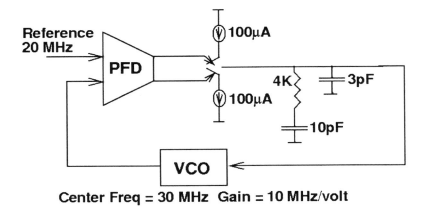

Figure 7.13. Simple charge-pump PLL

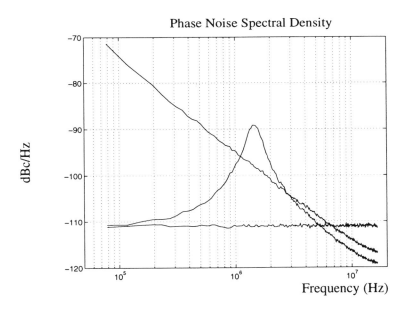

Figure 7.14. Phase noise spectrum: ideal PFD with noisy reference and VCO

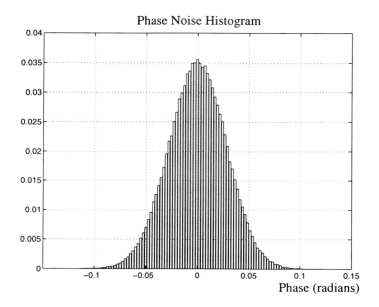

Figure 7.15. Phase noise histogram: ideal PFD with noisy reference and VCO

phase differences between the two inputs. To see the effect of the dead zone region on the resultant phase noise spectrum, we simulated the PLL circuit first with a noiseless reference and VCO, and a PFD which can not generate pulses at its outputs with a width less than 2 *nsecs*. We also performed a simulation with a white phase noise source at the reference, a random walk phase noise model for the VCO, and the same PFD with the dead zone. Figure 7.16 shows the spectrum of phase noise for the closed-loop VCO for both cases. Figure 7.16 is the histogram of the sampled phase noise for the case with a noisy reference and VCO, and a PFD with a dead zone. To eliminate the phase noise performance degradation caused by the dead zone of the PFD, an "alive zone" PFD can be used. The alive zone phase comparator introduces a reset delay into a standard edge-sensitive PFD and creates minimum duration pulses at both of its outputs. These occur at each phase comparison and activate the fourth state of the charge pump which has both positive and negative currents on at the same time, a state not possible for a standard PFD. The positive and negative charge pumps both deliver a charge greater than zero at each phase comparison. With zero phase error, ideally the net charge pumped sums to zero. However any mismatches between the positive and negative currents will cause phase noise performance degradation. To see the effect of the current

BEHAVIORAL MODELING AND SIMULATION OF PHASE-LOCKED LOOPS   255

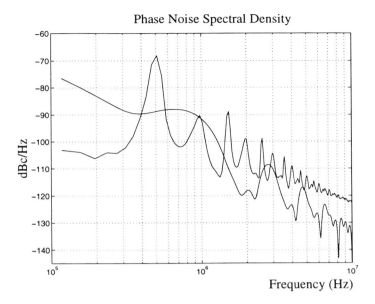

Figure 7.16. Phase noise spectrum: dead-zone PFD with noisy reference and VCO

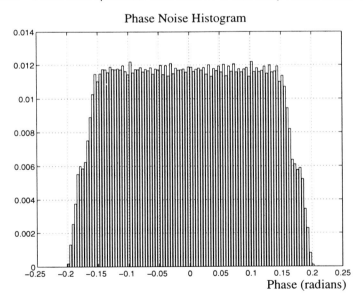

Figure 7.17. Phase noise histogram: dead-zone PFD with noisy reference and VCO

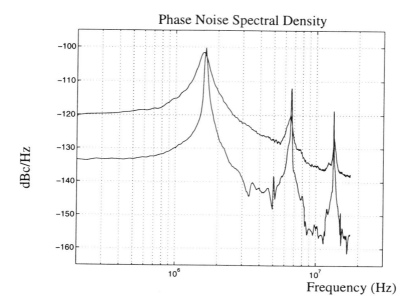

Figure 7.18. Phase noise spectrum: alive-zone PFD with noisy reference and VCO

mismatch, we simulated the PLL circuit first with a noiseless reference and VCO, an alive zone PFD with a minimum pulse duration of 3 $nsecs$, and a charge-pump with 10% current mismatch. We also performed a simulation with a white phase noise source at the reference, a random walk phase noise VCO, and the same alive zone PFD with a 10% current mismatch. Figure 7.18 shows the phase noise spectrum of the closed-loop VCO for both cases.

The phase noise spectrum obtained with behavioral simulation for the PLL of [64, 4] is shown in Figure 7.19 along with the phase noise spectrum of the open-loop VCO.

7.6 Summary

In this chapter, we presented a hierarchical behavioral modeling and simulation methodology for the design of phase-locked loops used in clock generation and frequency synthesis applications.

We first presented a simple LTI analysis of a typical PLL architecture. We used the results of this simple LTI analysis in creating the behavioral models of the PLL components, and making simplifications in the models which increased the efficiency of behavioral simulation. We then described the behavioral models of PLL components that capture various nonidealities including the phase

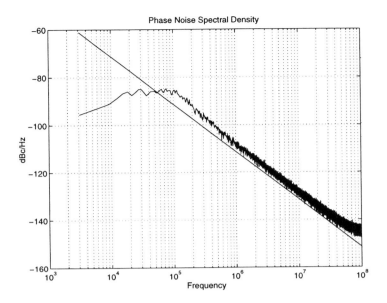

Figure 7.19. Phase noise spectrum: clock generator PLL

noise/timing jitter behavior. When developing the models, we heavily relied on the results we obtained in Chapter 6 for the phase noise behavior of free running oscillators.

The behavioral simulation of the PLL is done in time domain using an algorithm which has the following features:

- Numerical integration algorithm with threshold crossing detection and break points.

- PFD event queue scheduling and evaluation.

- Realization of the phase noise/timing jitter processes using a Gaussian random number generator. Phase noise/timing jitter sources are introduced into the time domain simulation as random threshold values. When certain state variables during the numerical integration cross these random threshold values, a transition occurs at the node represented by that state variable.

- Automatic acquisition detection.

The spurious tones and the timing jitter/phase noise of the output for the PLL that is in lock is characterized by postprocessing the data that is produced during behavioral simulation. In particular, the spurious tones are estimated by first performing Fourier analysis on the dumped control voltage for the VCO, and then by using FM theory to predict the magnitude of the tones at certain frequencies at the VCO output. The variance of the timing jitter of the transitions as well as the spectrum of the timing jitter/phase noise for the closed-loop VCO output is estimated by postprocessing the transition times dumped during behavioral simulation. A sample path for the timing jitter process is created using self-referencing. The variance of the timing jitter is calculated by first calculating the sample mean and then the sample variance for this sample path. The spectrum of phase noise/timing jitter is estimated by using the time-averaged periodogram technique on a sample path of the self-referenced timing jitter process.

We believe one of the important conclusions that can be drawn from this chapter is that behavioral modeling and simulation can be an invaluable tool in designing analog and mixed-signal systems. However, it is not possible to develop and use a behavioral modeling and simulation methodology for any type of system without a good understanding of its operation principles and various nonidealities that have a significant impact on the performance. Having a circuit simulator that comes with a hardware description language which has various constructs that help the system design engineer develop behavioral models and simulation techniques is certainly useful. However the hardware description language is *merely* a convenient environment to implement the behavioral models and simulation techniques which were developed. Much work remains to be done to develop good behavioral models for various analog and mixed-signal systems. A "good" behavioral model is the one that captures important nonidealities, such as nonlinearities and noise, in a rigorous way using the most appropriate mathematical constructs and techniques.

Notes

1. For instance, see [50], [57].

2. This model obviously does not capture the complete behavior of the PFD, but we are going to use this model only for qualitative analysis.

3. To be precise, $T_d(f) = -T_i(f)$. However θ_i, θ_d and θ_o represent *zero-mean* stochastic processes, moreover, θ_i and θ_d are assumed to be independent. Hence, only the *magnitudes* of the transfer functions are relevant.

4. Actually, amplified (by N) versions of the phase fluctuations of the reference, FD, and the PFD appear as phase fluctuations at the output. However, note that the signal frequency at the output is also N times the frequency of the reference, and the signal frequencies at the output of the FD and the inputs of the PFD.

5. We assume that the divide ratios L and M for the FDs outside of the feedback loop are set to 1 in the model of Figure 7.1.

6. This means that the frequency of the VCO can change instantaneously when the control voltage changes. Of course, this is not possible in practice. This model can be easily generalized by introducing state variables to represent the memory in the nonlinear map g that relates the VCO frequency to the control voltage at the LF output. Furthermore, these state variables can be included in the model of the LF to be discussed.

7. Recall that the phase noise of a free running oscillator can be modeled as a random walk process when only white noise sources are considered. This means that we are modeling the phase noise of the reference oscillator and the VCO considering thermal and shot noise sources only, and ignoring the $1/f$ noise sources.

8. The period of time PLL needs to be simulated for depends on the particular PLL architecture. Roughly, PLL needs to be simulated for a period of time that is several, i.e. 10, loop time constants long. Loop time constant is roughly estimated from the time it takes the PLL to lock onto the reference signal.

9. This waveform is obtained by setting the control voltage input of a stand-alone VCO circuit to $g^{-1}(Nf_{ref})$ and then performing a transistor-level simulation (e.g. transient analysis, shooting method or harmonic balance simulation can be used) to calculate the periodic steady-state for the oscillator.

10. Note that it would not be correct to replace Nf_{ref} with $f_{out}(t)$.

11. Please see the discussion in Section 6.7.

12. This is similar to what is done in timing jitter measurements with sampling oscilloscopes [38]. A transition of the oscillation waveform is designated as the reference or *trigger* transition, and a histogram of transition times is compiled at a specified distance from this transition.

13. For SPICE simulation, macromodels were used for the frequency dividers instead of their transistor-level implementations. Both of the simulations were performed on a DEC Alpha machine with a CPU that is a DEC Alpha chip 21164 with 250 MHz clock frequency, 4 Mb of cache, and a SPEC int_92 of 277.

14. With a Tektronix 11801B high-bandwidth digitizing oscilloscope.

8 CONCLUSIONS AND FUTURE WORK

We presented analysis, simulation and characterization techniques, and behavioral models for noise in nonlinear electronic circuits and systems, along with practical examples. We treated the problem within the framework of, and using techniques from, the probabilistic theory of stochastic processes and stochastic differential systems.

We proposed a novel time-domain algorithm for the simulation and complete second-order probabilistic characterization of the behavior of nonlinear electronic circuits in the presence of noise. With this algorithm, one can simulate a nonlinear dynamic circuit with electrical noise sources and arbitrary large-signal excitations by directly calculating the correlation matrix of the state variables of the system which are represented by nonstationary stochastic processes. This method enables us to analyze transient and nonstationary noise phenomena since a steady-state condition for the circuit is not required. We described the implementation of the noise simulation algorithm for numerical computation. The noise simulation algorithm is a core tool which can be used to investigate, simulate, understand, and model various noise phenomena in nonlinear analog and mixed-signal circuits.

We used the noise simulation algorithm to investigate, understand and model the phase noise/timing jitter phenomenon in oscillator circuits. We presented a formal definition for phase noise, proposed algorithms for its probabilistic characterization, and described their implementations for numerical computation. We also developed behavioral models for phase noise of free running oscillators.

We then presented a hierarchical behavioral specification and simulation methodology for the design of phase-locked loops used in clock generation and frequency synthesis applications. We developed behavioral models of phase-locked loop components which capture various nonidealities including the phase noise/timing jitter and spurious tone behavior. We presented a mixed-signal behavioral simulation algorithm and techniques to postprocess the data from behavioral simulation to characterize the spurious tones and the timing jitter/phase noise of the output of the phase-locked loop.

Much work remains to be done for the theory, analysis, and simulation of noise in nonlinear electronic circuits and systems. Even more work remains to be done in *fully* applying the techniques proposed by us and others in real design problems. We believe that one of the open problems is to form a *rigorous* connection between the theory of noise in nonlinear circuits and what one sees and measures in practice on a piece of equipment.

The time domain non-Monte Carlo noise simulation algorithm we proposed was intended for characterizing the behavior of "small" (i.e. with several hundred state variables) subblocks (e.g. analog blocks such as mixers and oscillators) of a mixed-signal system design. However, even small analog building blocks can be quite complicated resulting in a large number of equations for noise analysis. With the current numerical methods used in the noise simulation algorithm, the noise simulation of circuits with more than a few hundred nodes can become quite prohibitive in terms of CPU time. We believe that Krylov subspace based iterative techniques with a specific preconditioner designed for the special linear system of equations that needs to be solved at every time point is a promising avenue to explore in an attempt to reduce the computational cost of noise simulation.

Almost all of the noise analysis techniques proposed (by us and others) assume that the effect of the noise sources on the circuit is "small" and hence use a *linear* model of the system for noise analysis. The small noise assumption and the use of a linear model are not always justified. One usually has a qualitative "feeling" about the noise analysis problem in hand, and can give an intuitive argument about the validity of the small noise assumption. Ideally, we would like to have a *quantitative* measure to judge the validity of the small noise assumption, i.e. a rigorous probabilistic error analysis for the small noise expansion that results in a linear model.

The need for further theoretical and experimental work on the basic noise models for the electrical noise sources in semiconductor devices is well warranted. We believe that the work of Coram and Wyatt [22] is a pioneering one in this area. However, it needs to be complemented with work that will result in rigorously derived "approximate" models for basic noise phenomena, such as thermal noise in nonlinear dissipative devices. Then, this work should be followed by experimental work that will test the validity of the derived approximate noise models.

We believe that the current design techniques and methodologies for nonlinear analog and mixed-signal electronic systems can *not* fully utilize the information in the noise performance characterizations one can calculate using the nonlinear noise analysis techniques we proposed. The basic reason for this is that the design techniques currently used are mostly based on concepts from the theory of LTI systems and WSS noise sources. The connection between the noise performance characterizations of the building blocks of a system and the overall performance specifications needs to be rigorously studied. For instance, for an RF transceiver system, the key performance specification for the overall system is the *bit error rate* (BER). The rigorous connection between the BER for the whole system and the noise performance characterizations of the building blocks needs to be established. We believe that the currently used techniques to establish this connection are rather simplistic and ad-hoc. This probably results in a design with components that are over-specified to compensate for the simplistic analysis that establishes the connection between the overall system specifications and the specifications for the components that make up the system. We believe that a top-down hierarchical design methodology [4] is the most effective framework to establish the rigorous connection between the overall noise performance specifications, e.g. BER, and the noise performance characterizations of the components, e.g. the phase noise of the LO.

We already presented a hierarchical methodology for the noise performance modeling and characterizations of phase-locked loops used in clock generation and frequency synthesis applications. This hierarchical methodology should be extended for the noise performance characterizations of other building blocks of an RF transceiver system, e.g. the mixer, and eventually for a whole RF transceiver system. For phase-locked loops, we first developed a rigorous model for the noise performance characterization of the open-loop VCO, and then used this characterization in the analysis and characterization of the whole phase-locked loop circuit. Similarly, rigorous characterizations for various building blocks of an RF transceiver system should be first developed, and then these characterizations should be used in the analysis and characterization of the whole system. A behavioral specification and simulation methodology based on the Volterra/Wiener theory of nonlinear systems seems to be a promising approach for the analog front-end of RF transceiver systems, even though there are some disadvantages and issues to be worked out. For behavioral modeling and simulation of the analog front-end of RF systems, the key problem is the proper and rigorous modeling of nonlinear frequency dependent effects and also the nonlinear noise behavior of the components that make up the system. We believe that undertaking a real design example will be the most effective way to develop a hierarchical simulation, characterization and design methodologie for RF transceiver systems that can rigorously and fully utilize the techniques we proposed in this work. Such a hierarchical methodology will enable the RF system designer to invent new architectures and optimize a given architecture for power, area, etc. while it still meets the specifications.

References

[1] P.R. Gray and R.G. Meyer, *Analysis and Design of Analog Integrated Circuits*, John Wiley & Sons, second Edition, 1984.

[2] A. Demir, E. Liu and A. Sangiovanni-Vincentelli, "Time-Domain non-Monte Carlo Noise Simulation for Nonlinear Dynamic Circuits with Arbitrary Excitations", in *Proc. IEEE/ACM ICCAD*, November 1994.

[3] A. Demir, E. Liu and A. Sangiovanni-Vincentelli, "Time-Domain non-Monte Carlo Noise Simulation for Nonlinear Dynamic Circuits with Arbitrary Excitations", *IEEE Transactions on Computer-Aided Design of Integrated Circuits and Systems*, vol. 15, pp. 493–505, May 1996.

[4] H. Chang, E. Charbon, U. Choudhury, A. Demir, E. Felt, E. Liu, E. Malavasi, A. Sangiovanni-Vincentelli and I. Vassiliou, *A Top-Down Constraint-Driven Design Methodology for Analog Integrated Circuits*, Kluwer Academic Publishers, 1997.

[5] W.A. Gardner, *Introduction to Random Processes with Applications to Signals & Systems*, McGraw-Hill, second Edition, 1990.

[6] G.R. Grimmet and D.R. Stirzaker, *Probability and Random Processes*, Oxford Science Publications, second Edition, 1992.

[7] A. Papoulis, *Probability, random variables, and stochastic processes*, New York: McGraw-Hill, second Edition, 1984.

[8] L.A. Zadeh, "Frequency Analysis of Variable Networks", *Proceedings of the I.R.E.*, March 1950.

[9] L. Arnold, *Stochastic Differential Equations: Theory and Applications*, John Wiley & Sons, 1974.

[10] C.W. Gardiner, *Handbook of Stochastic Methods for Physics, Chemistry and the Natural Sciences*, Springer-Verlag, 1983.

[11] P.E. Kloeden and E. Platen, *Numerical Solution of Stochastic Differential Equations*, Springer-Verlag, Berlin, New York, 1992.

[12] V.S. Pugachev and I.N. Sinitsyn, *Stochastic Differential Systems: Analysis and Filtering*, Wiley, Chichester, Sussex, New York, 1987.

[13] K. Sobczyk, *Stochastic Differential Equations*, Kluwer Academic Publishers, 1991.

[14] H. D'Angelo, *Linear Time-Varying Systems: Analysis and Synthesis*, Allyn and Bacon, 1970.

[15] H. Haken, *Advanced Synergetics*, Springer-Verlag, 1983.

[16] F. X. Kaertner, "Analysis of White and $f^{-\alpha}$ Noise in Oscillators", *International Journal of Circuit Theory and Applications*, vol. 18, pp. 485–519, 1990.

[17] M.S. Gupta, "Thermal Noise in Nonlinear Resistive Devices and its Circuit Representation", *Proceedings of the IEEE*, vol. 70, n. 8, pp. 788, August 1982.

[18] F.N.H. Robinson, *Noise and Fluctuations in Electronic Devices and Circuits*, Clarendon Press, Oxford, 1974.

[19] R.A. Colclaser and S. Diehl-Nagle, *Materials and Devices for Electrical Engineers and Physicists*, McGraw-Hill, 1985.

[20] Y.P. Tsividis, *Operation and Modeling of the MOS Transistor*, McGraw-Hill, 1987.

[21] M.S. Keshner, "$1/f$ Noise", *Proceedings of the IEEE*, vol. 70, n. 3, pp. 212, March 1982.

[22] J. Wyatt and G. Coram, "Thermodynamic Properties of Simple Noise Models for Solid-State Devices", Technical report, MIT, 1996.

[23] B. Pellegrini, R. Saletti, B. Neri and P. Terreni, "$1/f^v$ Noise Generators", in *Noise in Physical Systems and 1/f Noise*, page 425, 1985.

[24] A. Ambrozy, *Electronic Noise*, McGraw-Hill, 1982.

[25] A. van der Ziel, *Noise in Solid-State Devices and Circuits*, John Wiley & Sons, 1986.

[26] R. Rohrer, L. Nagel, R.G. Meyer and L. Weber, "Computationally Efficient Electronic-Circuit Noise Calculations", *IEEE Journal of Solid-State Circuits*, vol. SC-6, n. 4, pp. 204, August 1971.

[27] J. Vlach and K. Singhal, *Computer Methods for Circuit Analysis and Design*, Van Nostrand Reinhold, second Edition, 1994.

[28] B. Johnson, T. Quarles, A.R. Newton, D.O. Pederson and A. Sangiovanni-Vincentelli, *SPICE3 Version 3e User's Manual*, University of California, Berkeley, April 1991.

[29] K.S. Kundert, J.K. White and A. Sangiovanni-Vincentelli, *Steady-State Methods for Simulating Analog and Microwave Circuits*, Kluwer Academic Publishers, 1990.

[30] M. Okumura, H. Tanimoto, T. Itakura and T. Sugawara, "Numerical Noise Analysis for Nonlinear Circuits with a Periodic Large Signal Excitation Including Cyclostationary Noise Sources", *IEEE Transactions on Circuits and Systems-1: Fundamental Theory and Applications*, vol. 40, n. 9, pp. 581, September 1993.

[31] R. Telichevesky, K.S. Kundert and J. White, "Efficient AC and Noise Analysis of Two-Tone RF Circuits", in *Proc. Design Automation Conference*, June 1996.

[32] C.D. Hull, *Analysis and Optimization of Monolithic RF Down Conversion Receivers*, PhD thesis, University of California, Berkeley, 1992.

[33] C.D. Hull and R.G. Meyer, "A Systematic Approach to the Analysis of Noise in Mixers", *IEEE Transactions on Circuits and Systems-1: Fundamental Theory and Applications*, vol. 40, pp. 909, December 1993.

[34] J.S. Roychowdhury and P. Feldmann, "Cyclostationary Noise Analysis of Nonlinear Circuits", Technical report, Bell Laboratories, Murray Hill, New Jersey, 1996.

[35] R. Melville, P. Feldmann and J. Roychowdhury, "Efficient Multi-tone Distortion Analysis of Analog Integrated Circuits", in *Proc. IEEE Custom Integrated Circuits Conference*, May 1995.

[36] R.W. Freund and P. Feldmann, "Efficient Small-Signal Circuit Analysis and Sensitivity Computations with the PVL Algorithm", in *Proc. ACM/IEEE ICCAD*, November 1994.

[37] P. Bolcato and R. Poujois, "A New Approach for Noise Simulation in Transient Analysis", in *Proc. IEEE International Symposium on Circuits & Systems*, May 1992.

[38] J.A. McNeill, *Jitter in Ring-Oscillators*, PhD thesis, Boston University, 1994.

[39] A.L. Sangiovanni-Vincentelli, "Circuit Simulation", in *Computer Design Aids for VLSI Circuits*, Sijthoff & Noordhoff, The Netherlands, 1980.

[40] Z. Gajic and M.T. Javed Qureshi, *Lyapunov Matrix Equation in System Stability and Control*, Academic Press, 1995.

[41] S.J. Hammarling, "Numerical Solution of the Stable, Non-negative Definite Lyapunov Equation", *IMA Journal of Numerical Analysis*, vol. 2, pp. 303–323, 1982.

[42] C. Pommerell and W. Fichtner, "PILS: An Iterative Linear Solver for Ill-Conditioned Systems", in *Supercomputing*, pp. 588–599, 1991.

[43] E. Tomacruz, J. Sanghavi and A. Sangiovanni-Vincentelli, "A Parallel Iterative Linear Solver for Solving Irregular Grid Semiconductor Device Matrices", in *Supercomputing*, 1994.

[44] R.H. Bartels and G.W. Stewart, "Solution of the Equation $AX + XB = C$", *Commun. ACM*, vol. 15, pp. 820–826, 1972.

[45] G.H Golub, S. Nash and C. Van Loan, "A Hessenberg-Schur Method for the Problem $AX + XB = C$", *IEEE Transactions on Automatic Control*, vol. AC-24, n. 6, pp. 909, December 1979.

[46] A.S. Hodel and S.T. Hung, "Solution and Applications of the Lyapunov Equation for Control Systems", *IEEE Transactions on Industrial Electronics*, vol. 39, n. 3, pp. 194, June 1992.

[47] Y. Saad, "Numerical Solution of Large Lyapunov Equations", in *Signal Processing, Scattering and Operator Theory, and Numerical Methods*, pp. 503–511, 1990.

[48] I. Jaimoukha and E. Kasenally, "Krylov Subspace Methods for Solving Large Lyapunov Equations", *SIAM Journal of Numerical Analysis*, vol. 31, pp. 227–251, 1994.

[49] T.L. Quarles, *Analysis of Performance and Convergence Issues for Circuit Simulation*, PhD thesis, University of California, Berkeley, 1989.

[50] F.M. Gardner, *Phaselock Techniques*, Wiley, 1979.

[51] T.C. Weigandt, B. Kim and P.R. Gray, "Analysis of Timing Jitter in CMOS Ring-Oscillators", in *Proc. IEEE ISCAS*, June 1994.

[52] A.A. Abidi and R.G. Meyer, "Noise in Relaxation Oscillators", *IEEE Journal of Solid-State Circuits*, December 1983.

[53] T.J. Aprille and T.N. Trick, "Steady-state Analysis of Nonlinear Circuits with Periodic Inputs", *Proceedings of the IEEE*, vol. 60, n. 1, pp. 108–114, Jaunary 1972.

[54] T.J. Aprille and T.N. Trick, "A Computer Algorithm to Determine the Steady-state Response of Nonlinear Oscillators", *IEEE Transactions on Circuit Theory*, vol. CT-19, n. 4, pp. 354–360, July 1972.

[55] R. Telichevesky, K.S. Kundert and J. White, "Efficient Steady-State Analysis Based on Matrix-Free Krylov-Subspace Methods", in *Proc. Design Automation Conference*, June 1995.

[56] D.B. Leeson, "A Simple Model of Feedback Oscillator Noise Spectrum", *Proceedings of the IEEE*, vol. 54, n. 2, pp. 329, February 1966.

[57] U.L. Rohde, *Digital PLL Frequency Synthesizers: Theory and Design*, Prentice-Hall, 1983.

[58] M. Farkas, *Periodic Motions*, Springer-Verlag, 1994.

[59] F. X. Kaertner, "Determination of the Correlation Spectrum of Oscillators with Low Noise", *IEEE Transactions on Microwave Theory and Techniques*, vol. 37, n. 1, pp. 90–101, January 1989.

[60] E. Liu, *Analog Behavioral Simulation and Modeling*, PhD thesis, University of California, Berkeley, 1993.

[61] E. Liu and A. Sangiovanni-Vincentelli, "Behavioral Representations for VCO and Detectors in Phase-Lock Systems", in *Proc. IEEE Custom Integrated Circuits Conference*, May 1992.

[62] D.R. Preslar and J.F. Siwinski, "An ECL/I^2L Frequency Synthesizer for AM/FM Radio with an Alive Zone Phase Comparator", *IEEE Transactions on Consumer Electronics*, August 1981.

[63] K.S. Kundert, *The Designer's Guide to SPICE and SPECTRE*, Kluwer Academic Publishers, 1995.

[64] I. Vassiliou, H. Chang, A. Demir, E. Charbon, P. Miliozzi and A. Sangiovanni-Vincentelli, "A Video Driver System Designed Using a Top-Down Constraint-Driven Design Methodology", in *Proc. IEEE/ACM ICCAD*, November 1996.

About the Authors

Alper Demir

Alper Demir was born in Konya, Turkey, in 1969. He received the B.S. degree in electrical engineering from Bilkent University, Turkey, the M.S. and the Ph.D. degrees in electrical engineering and computer sciences from the University of California, Berkeley in 1991, 1994 and 1997 respectively.

From May 1992 to January 1997 he worked as a research and teaching assistant in the Electronics Research Laboratory and the EECS department at the University of California, Berkeley. He was with Motorola, Inc. during Summer 1995, and with Cadence Design Systems during Summer 1996. Dr. Demir joined Bell Laboratories (Lucent Technologies) as a member of the technical staff in January 1997.

In 1991, Alper Demir received the Regents Fellowship from the University of California at Berkeley, and was selected to be an Honorary Fellow of the Scientific and Technical Research Council of Turkey (TÜBİTAK).

Alberto Sangiovanni-Vincentelli

Alberto Sangiovanni Vincentelli was born in June 23rd, 1947 in Milan, Italy. He obtained a Doctor of Engineering Degree in Electrical Engineering and Computer Science from the Politecnico di Milano in 1971. From 1971 to 1976 he was a "Ricercatore" and a "Professore Incaricato" at the Politecnico di Milano. He joined the Department of Electrical Engineering and Computer Science of the University of California at Berkeley in 1976. He was Visiting Scientist at the IBM T.J. Watson Research Center in 1980, Consultant in residence at Harris in 1981 and Visiting Professor at MIT in 1987. He has held a number of visiting professor positions at the University of Torino, University of Bologna, University of Pavia, University of Pisa and University of Rome. He is a member of the Berkeley Roundtable for International Economy (BRIE) and a member of the Advisory Board of the Lester Center of the School of Business Administration.

He has been active in the high-tech industry for the past 15 years. He is a co-founder of Cadence and Synopsys, the two leading companies in the area of Electronic Design Automation. He is a corporate Fellow at Harris, he was a Director of ViewLogic and Pie Design System and Chair of the Technical Advisory Board of Synopsys. He is on the Board of Directors of Cadence, the Chair of its Scientific Advisory Council, the founder and a member of the Management Board of the Cadence Berkeley Laboratories. He consulted for several major US, (including AT&T, IBM, DEC, GTE, Intel, NYNEX, General Electric, Texas Instruments),

European (Alcatel, Bull, Olivetti, SGS-Thomson, Fiat) and Japanese companies (Fujitsu, Hitachi and Kawasaki Steel), in addition to start-ups (Actel, CrossCheck, Biocad, Redwood Design Automation, SiArc, Wireless Access). He was the Technology Adviser to Greylock Management. He is on the International Advisory Board of the Institute for Micro-Electronics of Singapore.

Alberto Sangiovanni-Vincentelli has been a contributor to the field of Computer-Aided Design of Electronic Systems. In particular, he has been active in simulation, synthesis and optimization of large digital and analog systems, parallel computing, embedded systems design, machine learning and discrete event systems.

In 1981 he received the Distinguished Teaching Award of the University of California. He has received the 1995 Graduate Teaching Award of the IEEE. He has received three Best Paper Awards (1982, 1983 and 1990) and a Best Presentation Award (1982) at the Design Automation Conference and is the co-recipient of the Guillemin-Cauer Award (1982-1983), the Darlington Award (1987-1988) and the Best Paper Award of the Circuits and Systems Society of the IEEE (1989-1990). He has published over 350 papers and five books in the area of design methodologies and tools. Dr. Sangiovanni-Vincentelli is a fellow of the Institute of Electrical and Electronics Engineers and was the Technical Program Chair of the International Conference on CAD for 1989 and the General Chair for 1990.

Index

1/f noise, 73
σ-field, 6
$\mathcal{F}$-measurable, 7
Behavioral modeling, 216
  charge pump, 230
  frequency divider, 226
    timing jitter, 227
  loop filter, 230
  phase-frequency detector, 228
    alive-zone, 230
    dead-zone, 229
  reference oscillator, 224
  VCO, 224
Behavioral simulation
  phase-locked loops
    algorithm, 233
    phase noise, 240
    spurios tones, 238
    timing jitter, 240
Central limit theorem, 12
Conditional probability, 7
Covariance matrix, 11
Differential equations, 42
  linear homogeneous, 43
  linear inhomogeneous, 44
  linear
    constant coefficients, 44
    Floquet theory, 45
    fundamental matrix, 43
    periodic coefficients, 45
    state transition matrix, 43
  Peano-Picard-Lindelöf iteration, 43
  uniqueness and existence of solutions, 42
Distribution function, 7
  conditional, 9
  joint, 7
Dynamical systems
  representation, 31
  representation
    linear systems, 32
Electrical noise, 68
Empirical autocorrelation, 28
Ensemble average, 14
Ergodicity, 26
  mean-square, 28
Event, 6
  independent, 7
Expectation, 9
Exponential distribution, 8
FFT, 30
Flicker noise, 73
Floquet theory, 45
Fokker-Planck equation, 62, 118
Gaussian distribution, 8
Lack of memory, 8
Langevin's equation, 50
Law of large numbers, 12
Lyapunov matrix equation
  algebraic, 139
  algebraic
    numerical solution, 139
  differential, 135
    numerical solution, 136
Matrices, 38
  function of, 40
  similar, 39
MNA, 100
Modified nodal analysis, 100

Noise analysis
  adjoint network, 105
  autocorrelation matrix, 131
    ODE for, 132
  closure approximations, 119
  examples
    mixer, 150
    negative resistance oscillator, 156
    parallel RLC circuit, 148
    switching noise of an inverter, 150
  formulation, 114
  linear time-varying SDE model, 125
  LPTV, 106
  nonstationary, 113
  periodic steady-state, 146
  perturbation approximations, 120
  probabilistic characterization, 118
  time-domain non-Monte Carlo, 113
  time-invariant steady-state, 145
  time-varying moments, 118
  time-varying variance-covariance matrix, 130
    ODE for, 130
Normal distribution, 8
Periodogram, 29
Phase noise
  concept, 165
  definition, 170
  due to correlated noise source, 201
  model as a random walk process, 174
  numerical characterization
    algorithm, 175, 192
    examples, 176, 195
  phase-locked loop, 209
  probabilistic characterization, 172
Phase-locked loops
  acquisition behavior, 247
  charge pump, 219
  LTI phase domain analysis, 219
  phase noise
    histogram, 254
  timing jitter
    examples, 249
Poisson distribution, 8
Positive definite/semidefinite, 41
Probability density function, 8
  conditional, 9
Probability mass function, 8
Probability measure, 6
Probability space, 7
Random variable, 7
  central moment, 10

  convergence of sequence of, 11
    almost surely, 11
    in distribution, 11
    in probability, 11
    in the rth mean, 11
  covariance, 10
  independent, 9
  mean, 10
  moment, 10
  sequence of, 11
  variance, 10
Random walk, 174
Sample space, 6
Shot noise
  model as a stochastic process, 74
  physical origin, 70, 72
Statistical mechanics, 68
Stochastic differential equations, 58
  numerical solution, 63
Stochastic differential systems, 49
Stochastic differentials
  Ito's formula, 60
Stochastic integral, 55
  Ito vs. Stratonovich, 61
Stochastic perturbation, 120
Stochastic process, 13
  as input to LPTV system, 37
  as input to LTI system, 35
  autocorrelation, 14
  continuity, 22
  cyclostationary, 17
  differentiability, 22
  ergodic, 27
  Gaussian, 15
  Markov, 16
  numerical simulation, 30
  Poisson, 21
  regular, 27
  sample path, 13
  spectral correlation, 20
  spectral density, 18
    time-varying, 19
  strictly stationary, 16
  white noise, 24
  wide-sense stationary, 16
  Wiener, 20
Thermal noise
  model as a stochastic process, 83
  Nyquist's theorem, 70
  physical origin, 69
Thermodynamic equilibrium, 68

Time averages, 14
Timing jitter
  concept, 165
  definition, 170
  probabilistic characterization, 172
Volterra/Wiener theory, 264
White noise, 24